Advances in Analytics for Learning and Teaching

Series Editors
Dirk Ifenthaler
Learning, Design and Technology
University of Mannheim
Mannheim, Baden-Württemberg, Germany

David C. Gibson
Teaching and Learning
Curtin University
Bentley, WA, Australia

This book series highlights the latest developments of analytics for learning and teaching as well as providing an arena for the further development of this rapidly developing field.
It provides insight into the emerging paradigms, frameworks, methods, and processes of managing change to better facilitate organizational transformation toward implementation of educational data mining and learning analytics. The series accepts monographs and edited volumes focusing on the above-mentioned scope, and covers a number of subjects. Titles in the series *Advances in Analytics for Learning and Teaching* look at education from K-12 through higher education, as well as vocational, business, and health education. The series also is interested in teaching, learning, and instructional design and organization as well as data analytics and technology adoption.

Vitomir Kovanovic • Roger Azevedo
David C. Gibson • Dirk lfenthaler
Editors

Unobtrusive Observations of Learning in Digital Environments

Examining Behavior, Cognition, Emotion, Metacognition and Social Processes Using Learning Analytics

Springer

Editors
Vitomir Kovanovic
Centre for Change and Complexity
University of South Australia
Adelaide, SA, Australia

Roger Azevedo
School of Modeling Simulation
and Training
University of Central Florida
Orlando, FL, USA

David C. Gibson
Data Science in Higher Education Learning
and Teaching, Curtin University
Bentley, WA, Australia

Dirk lfenthaler
Learning, Design and Technology
University of Mannheim
Mannheim, BW, Germany

Data Science in Higher Education
Learning and Teaching
Curtin University
Bentley, WA, Australia

ISSN 2662-2122 ISSN 2662-2130 (electronic)
Advances in Analytics for Learning and Teaching
ISBN 978-3-031-30991-5 ISBN 978-3-031-30992-2 (eBook)
https://doi.org/10.1007/978-3-031-30992-2

This Springer imprint is published by the registered company Springer Nature Switzerland AG
The registered company address is: Gewerbestrasse 11, 6330 Cham, Switzerland

Preface

The design of digital learning environments involves the idea of using computers for supporting human reasoning and learning processes – an old dream of artificial intelligence. Such applications are thought to be designed to execute operations of logical thinking using a multitude of rules which express logical relationships between terms and data in the Internet. In view of the countless unfulfilled promises of artificial intelligence in the 1980s and 1990s, however, one would be well advised to remain skeptical on this point.

More recently, emerging foundations of theory and analysis based on observation of digital traces have been enhanced by data science, particularly machine learning, with extensions to deep learning, natural language processing and artificial intelligence. These unobtrusive observation innovations have been brought into service to better understand higher-order thinking capacities such as self-regulation, collaborative problem-solving and the social construction of knowledge.

This edited volume presents a collection of articles concerning indicators or measurements of learning processes and related behavior, metacognition, emotion and motivation, as well as social processes. In addition, the book includes invited commentaries from a related field, such as educational psychology or cognitive science.

In *Unobtrusive Observations of Learning in Digital Environments*, we hope to advance the literature on artificial intelligence in education and add to the foundations of unobtrusive measurement. It features two major parts: Part I – Learning Processes, and Part II – Learning Data.

The editors are grateful for the assistance of experts in the field of artificial intelligence and education, who helped prepare this volume for publication. We also wish to thank our board of reviewers for their role in reviewing and editing the chapters.

Adelaide, SA, Australia — Vitomir Kovanovic
Orlando, FL, USA — Roger Azevedo
Bentley, WA, Australia — David C. Gibson
Mannheim, BW, Germany — Dirk Ifenthaler

Preface

Contents

About the Editors

Vitomir Kovanovic (Vitomir.Kovanovic@unisa.edu.au) is the Senior Lecturer in Learning Analytics at the Centre for Change and Complexity in Learning (C3L), Education Futures, University of South Australia, Australia. His research focuses on learning analytics within high school and university settings, looking at student self-regulation and study strategies. Vitomir is an Associate Editor for the *Higher Education Research & Development* journal (Taylor and Francis) and Academic Editor for *PLoS ONE* Journal (Public Library of Science). He was also Program Chair for the 2020 Learning Analytics & Knowledge Conference (LAK'20).

Roger Azevedo (roger.azevedo@ucf.edu) is a Professor at the School of Modeling Simulation and Training, University of Central Florida (UCF). He is the Lead Scientist for UCF's Learning Sciences Faculty Cluster Initiative. His main research area includes examining the role of cognitive, metacognitive, affective, and motivational self-regulatory processes during learning with advanced learning technologies. He is the former editor of the *Metacognition and Learning* journal, a fellow of the American Psychological Association, and the recipient of the prestigious Early Faculty Career Award from the National Science Foundation.

David C. Gibson (David.C.Gibson@curtin.edu.au) is a Professor and UNESCO Chair on Data Science in Higher Education Learning and Teaching at Curtin University, Australia. His foundational research demonstrated the feasibility of bridging from qualitative information to quantifiable dynamic relationships in complex models that verify trajectories of organizational change. He provides thought leadership as a researcher, professor, learning scientist, and innovator. He is the creator of simSchool, a classroom flight simulator for preparing educators, and eFolio, an online performance-based assessment system, and provides vision and sponsorship for Curtin University's Challenge, a mobile, game-based learning platform.

Dirk Ifenthaler (dirk@ifenthaler.info) is a Professor and Chair of Economic and Business Education – Learning, Design and Technology at University of Mannheim, Germany, and UNESCO Deputy Chair on Data Science in Higher Education Learning and Teaching at Curtin University, Australia. His research outcomes include numerous co-authored books, book series, book chapters, journal articles, and international conference papers, as well as successful grant funding in Australia, Germany, and USA. He is the Editor-in-Chief of the *Technology, Knowledge and Learning* and Senior Editor of the *Journal of Applied Research in Higher Education.*

Part I
Learning Processes

Chapter 1
Unobtrusive Observations of Learning Processes

Vitomir Kovanovic, Roger Azevedo, David C. Gibson, and Dirk Ifenthaler

Abstract In this section, we have gathered articles that deal with the unobtrusive observation of learning processes. By 'unobtrusive observations', we mean a process of detecting and analysing features of learning that can be found in digital traces of someone's interaction with a designed digital experience. The experience might have been designed as an experiment or for learning, such as online learning in a massively open online course or an in-class exercise utilizing technology. By 'learning processes', we refer to various aspects of how someone interacts with the designed digital learning experience, including the emotions, self-regulation skills, problem-solving approaches, collaborative capabilities, and motivations of the learner. These aspects of learning are sometimes referred to as noncognitive, although a case can be made that all thinking, acting and emotional states have cognitive components. Higher-order constructs such as self-regulation, leadership, and collaboration are thought to be composed of, or clustered with, a more complex layering of underlying capabilities, like how individual letter recognition is part of reading for understanding.

V. Kovanovic (✉)
Centre for Change and Complexity, University of South Australia, Adelaide, SA, Australia
e-mail: Vitomir.Kovanovic@unisa.edu.au

R. Azevedo
School of Modeling Simulation and Training, University of Central Florida, Orlando, FL, USA
e-mail: roger.azevedo@ucf.edu

D. C. Gibson
Data Science in Higher Education Learning and Teaching, Curtin University, Bentley, WA, Australia
e-mail: david.c.gibson@curtin.edu.au

D. Ifenthaler
Data Science in Higher Education Learning and Teaching, Curtin University, Bentley, WA, Australia

Learning, Design and Technology, University of Mannheim, Mannheim, BW, Germany
e-mail: dirk@ifenthaler.info

V. Kovanovic et al. (eds.), *Unobtrusive Observations of Learning in Digital Environments*, Advances in Analytics for Learning and Teaching,
https://doi.org/10.1007/978-3-031-30992-2_1

Keywords Trace data · Multimodal data · Unobtrusive observation · Learning process

1 Section Overview

To make an unobtrusive observation requires a quiet detection of features, a detection that does not disturb the natural actions of the interacting learner. For example, a sensor system might be collecting near-real-time data about someone's physiological states during some task or activity while, at the same time, also collecting information about the tools used or communications with team members. Some of these detected features are then combined into indicators of states (e.g. engagement, deliberate pause) or trajectories (e.g. increased skill, changes in emotional valence) of interest. Along with the primary features and indicators, an observation also requires a bounded context, a surrounding set of nodes labelled as entities and edges labelled as relationships or processes if the analysis uses a network model.

Regarding the unobtrusive data collection about learning processes, both features and their context need to be engineered, which entails answering some critical questions. How are the indicators combined into features? What is the role of the extracted features in the learning process? How does the learner's awareness of the features and indicators impact their performance? What are the limitations of the affordances in the designed experience to elicit evidence of the constructs of interest? Added to this are a host of potential noncognitive influences like the emotional states, motivations, and social capital of facing a variety of learning tasks as a team member. In the following chapters, you will find discussions of features such as:

- Emotion, including emotional variability, instability, inertia, cross-lags, and emotional patterns (Chap. 2)
- Problem-solving, e.g. deliberate pause (Chap. 3)
- Soft skills, e.g. leadership skills in a workplace learning context (Chap. 4)
- Motivation, particularly that changes over time and entails changing contexts that require thinking about ongoing feature redefinition (Chap. 5)
- Self-regulated learning (Chap. 6)

So, the picture that is developing for unobtrusive observation is one that is both dynamic and contextual that requires multiple and wide-ranging measurements over time. Several data challenges arise, including dealing with differences in measures per minute and quality of the measures and aggregations from sensors collected using different time windows. Data must be integrated and clustered meaningfully to link to the indicators, a process that, at this time, requires both human and machine learning techniques. Understanding dynamic context requires a complex system perspective, for example, to determine the unit of analysis, the surrounding context, and the influences on the dynamics from the surround as well as how the unit of analysis influences its surround.

The following brief introductions provide a quick glimpse of how these authors view the unobtrusive observation of learning processes.

Chapter 2 Juan Zheng, Shan Li, Susanne P. Lajoie. *A Review of Measurements and Techniques to Study Emotion Dynamics in Learning*

Emotion states are dynamic and contextual across learning environments. Learners who experience similar levels of emotions can differ substantially in the fluctuation of emotions in a task or throughout a course. The authors introduce a taxonomy of emotion dynamics features, i.e. emotional variability, emotional instability, emotional inertia, emotional cross-lags, and emotional patterns. They discuss emotion detection methods that can unobtrusively capture longitudinal and time-series data, including experience sampling methods, emote aloud, facial expressions, vocal expressions, language and discourse, and physiological sensors. They also present several emerging techniques for assessing emotion dynamics, including entropy analysis, growth curve modelling, time series analysis, network analysis, recurrence quantification analysis, and sequential pattern mining.

Chapter 3 Karen D. Wang, Shima Salehi, Carl Wieman. *Applying Log Data Analytics to Measure Problem Solving in Simulation-Based Learning Environments*

This chapter presents the research team's efforts towards understanding how the log data of students' interactions within an educational simulation can be translated into meaningful evidence about their problem-solving process. Features extracted from log data were found to be both significant predictors of students' problem-solving outcomes and indicators of specific problem-solving practices. Deliberate pauses during the problem-solving process, in particular, were identified as an important and generalizable feature associated with problem-solving competencies across different tasks.

Chapter 4 Abhinava Barthakur, Vitomir Kovanovic, Srecko Joksimovic, Abelardo Pardo. *Challenges in Assessments of Soft Skills: Towards Unobtrusive Approaches to Measuring Student Success*

This chapter outlines a multi-tiered case study that used a novel blended methodology, marrying measurement models and learning analytics techniques, to mitigate some of the challenges of unobtrusively measuring leadership skills in a workplace learning context. Using learners' reflection assessments, several leadership-defining course objectives were quantified using a blend of natural language and structured data approaches. Student progress was assessed over time in relation to course learning outcomes. The chapter discusses the implications of their evidence-based assessment approach, informed by theory, to measure and model soft skills acquisition.

Chapter 5 Heeryung Choi, Philip H. Winne, Christopher Brooks. *Proposal and Critiques of Measuring Motivational Constructs Using State-Revealing Trace Data*

This chapter examines opportunities afforded by trace data to capture dynamically changing latent states and trajectories spanning states in self-regulated learning (SRL). The authors catalogue and analyse major challenges in temporally investigating SRL constructs related to a prominent motivational factor, achievement goals. The chapter summarizes three recent studies addressing these challenges and characterizes learning analytics designed to promote SRL and motivation formed from unobtrusive traces. The authors propose a research agenda for learning analytics focusing on guiding and supporting SRL.

Chapter 6 Sambit Praharaj, Maren Scheffel, Marcus Specht, Hendrik Drachsler. *Measuring Collaboration Quality Through Audio Data and Learning Analytics*

This chapter addresses the unobtrusive detection and measurement of collaboration quality based on audio recordings of student interactions. Using two indicators, time and content of communications, the team aimed to move towards an automated measure of collaboration quality. The authors explain the design of a sensor-based automatic analysis system and show their analysis using meaningful visualizations to gain insights into the quality of student collaboration.

To summarize, the detection methods discussed in the section include latent variable detection (Chap. 2), log traces becoming semantically meaningful units of analysis (Chap. 3), automated content analysis of learners' reflection assessments (Chap. 4), and sensors systems and data handling of noisy information (Chap. 6).

Analysis methods discussed in the chapters include entropy analysis, growth curve modelling, time series analysis, network analysis, recurrence quantification analysis, sequential pattern mining, quantitative association rule mining, cognitive diagnostic model machine scoring of natural language products for depth of reflection on leadership skills, and temporal challenges of dynamic and contextual data, to name a few.

As noted by these authors, the path from unobtrusively acquiring log data to analysing semantically meaningful evidence of learning processes is an interdisciplinary effort that joins personality psychology, developmental science, learning science, and neuroscience. We trust that you will find this collection useful.

Chapter 2
A Review of Measurements and Techniques to Study Emotion Dynamics in Learning

Juan Zheng, Shan Li, and Susanne P. Lajoie

Abstract Emotion states are dynamic and contextual across learning environments. Learners who experience similar levels of emotions can differ substantially in the fluctuation of emotions in a task or throughout a course. However, research on emotion dynamics is still limited and fragmented in teaching and learning contexts. Despite an increasing interest from researchers to investigate the dynamic aspect of students' emotions, there has been no review of measurements and techniques to study emotion dynamics. We address this gap by introducing a taxonomy of emotion dynamics features, i.e., emotional variability, emotional instability, emotional inertia, emotional cross-lags, and emotional patterns. Furthermore, we synthesize the current emotion detection methods that can unobtrusively capture longitudinal and time series data of emotions. These methods include experience sampling methods, emote-aloud, facial expressions, vocal expressions, language and discourse, and physiological sensors. Moreover, this review introduces the predominant analytical techniques that can quantify emotion dynamics from longitudinal and time series data. We demonstrate how the conventional statistical methods have been used to quantify different features of emotion dynamics. We also present some emerging techniques for assessing emotion dynamics, including entropy analysis, growth curve modeling, time series analysis, network analysis, recurrence quantification analysis, and sequential pattern mining. The emotion detection and analytical approaches described in this chapter provide researchers a practical guide in examining emotion dynamics in teaching and learning contexts. This chapter also has theoretical importance since it will help researchers develop a dynamic perspective of emotions and will promote a deep understanding of emotion generation and regulation.

J. Zheng (✉) · S. Li
Lehigh University, Bethlehem, PA, USA
e-mail: juz322@lehigh.edu; shla22@lehigh.edu

S. P. Lajoie
McGill University, Montreal, QC, Canada
e-mail: susanne.lajoie@mcgill.ca

V. Kovanovic et al. (eds.), *Unobtrusive Observations of Learning in Digital Environments*, Advances in Analytics for Learning and Teaching,
https://doi.org/10.1007/978-3-031-30992-2_2

Keywords Emotion dynamics · Emotional fluctuations · Emotion measurement · Analytical techniques · Learning

1 Introduction

A consensus is emerging that emotions play a critical role in students' learning and problem-solving (Gross, 2013; Lajoie et al., 2019; Pekrun, 2006; Pekrun et al., 2002; Schutz & Davis, 2000; Zheng et al., 2021). In fact, emotion-related studies are a growing feature in the landscape of educational research. Great effort has been made to understand how the features of emotions, such as the category (e.g., achievement and epistemic emotion), duration, intensity, valence (i.e., positive/negative), and arousal (i.e., activation/deactivation) of emotions, influence students' learning processes and outcomes directly or indirectly. However, it is noteworthy that emotion states are essentially dynamic and contextual across a range of learning environments. Learners who experience similar levels of emotions can differ substantially in the fluctuation of emotions in a task or throughout a course (Reitsema et al., 2022). The literature is still fragmented and limited regarding the dynamical features of emotions. For instance, emotion dynamics can be quantified as the variability, instability, or inertia of emotions (Houben et al., 2015). To our knowledge, those features have rarely been investigated in educational studies. The purpose of this chapter is to advance this field of study by presenting a review of measurements and techniques for researching emotion dynamics.

In this chapter, we will focus primarily on learners' emotion dynamics in teaching and learning contexts. As pointed out by Sperry et al. (2020), the variability, instability, and inertia aspects of emotion dynamics are extensively studied within the field of psychopathology. For instance, affective instability typically refers to a psychological illness related to emotional or affective dysregulation (Marwaha et al., 2014). In a systematic review of the literature on affective instability, Marwaha et al. (2014) defined it as "rapid oscillations of intense affect, with a difficulty in regulating these oscillations or their behavioral consequences" (p. 1082). A well-established measurement of affect lability is the Affective Lability Scale (ALS), which measures the affective changes between euthymia, depression, anxiety, anger, and hypomania (Oliver & Simons, 2004). In contrast to the flourishing of research on emotion dynamics in the psychiatric literature is the lack of attention to the dynamical features of emotions in teaching and learning contexts. Therefore, the aim of this chapter is to inform the study of emotion dynamics in academic learning and achievement settings by extracting insights from all available literature. This will facilitate our understanding of the components, features, and measurements of emotion dynamics that occur in student learning and problem-solving processes, laying a good foundation for future research.

Moreover, it is not hard to find that the literature on emotion, affect, and mood is extremely complex, given that researchers interchangeably use the terms emotion, affect, and mood for their studies. For instance, affective instability is often used interchangeably with affective lability, emotional instability, emotional lability, mood instability, and mood lability (Marwaha et al., 2014). As another example, researchers typically do not differentiate emotional variability from emotion, affect, or affective variability. In this chapter, we consider affect as a superordinate term for emotion and mood. In line with the modal model of emotion (Gross, 2013), emotions "involve person-situation transactions that compel attention, have meaning to an individual in light of currently active goals, and give rise to coordinated yet flexible multisystem responses that modify the ongoing person-situation transaction in crucial ways" (p. 5). To put it simply, emotions are intense, short-term responses to a contextual stimulus, yielding subjective experience, expressive behaviors, and cognitive, motivational, and physiological activations (Pekrun, 2006). In terms of mood, we deem it as a less intense affective state that lasts longer than an emotion, and it does not necessarily relate to a stimulus. In sum, this chapter distinguishes between the terms emotion, affect, and mood, to ease the conceptual complexity and to maintain a clear focus on learners' emotions. Thus, the terms emotional variability, emotional instability, and emotional inertia will be used throughout this chapter.

Furthermore, this chapter will focus on short-term dynamics of moment-to-moment emotions. As pointed out by Houben et al. (2015), historically, the studies on the features of emotion dynamics attempted to describe a person's emotional life, regardless of internal and external stimuli or conditions. In contrast, we are interested in the micro-level emotion dynamics that occur in specific learning and problem-solving contexts in a certain time period. This choice is made for the sake of assisting future researchers to investigate the mechanisms of how the various factors (e.g., prior knowledge, cognition, metacognition, motivation, learning environment, and task features) influence emotion dynamics. Only in this way we can hope to design effective interventions or scaffoldings to support students' learning. Therefore, the measurements for collecting emotion dynamics discussed in this chapter are mostly suitable for short-term dynamics of emotions. In the following sections, we first discuss the features of emotion dynamics. We then provide a review of prevalent measurement methods for collecting emotion dynamics, followed by an introduction to the most prominent techniques for analyzing emotion dynamics. Afterward, we list several challenges of studying emotion dynamics in learning. We close the chapter with a discussion of directions for future research.

2 The Features of Emotion Dynamics

There are various "elementary properties" of emotion dynamics (Krone et al., 2017). The most-studied properties of emotion dynamics, also known as emotion dynamic features (EDFs), are emotional variability, emotional instability, and emotional inertia (Houben et al., 2015; Sperry et al., 2020). Kuppens and Verduyn (2015) further proposed that EDFs could be organized into four categories, i.e., emotional variability, emotional covariation, emotional inertia, and emotional cross-lags. To provide a synthesis of the literature on defining emotion dynamics, we introduce a taxonomy of EDFs that consists of five essential features: emotional variability, emotional instability, emotional inertia, emotional cross-lags, and emotional patterns. It is noteworthy that we consider emotional covariation, which describes the co-occurrences of multiple emotions across time, as a type of emotional pattern.

2.1 *Emotional Variability*

Perhaps the most straightforward definition of emotional variability is "the extent to which the intensity of an emotion as experienced by an individual varies across time" (Krone et al., 2017, p. 740). To quantify emotional variability, an overall emotional score is typically obtained multiple times, whereby the within-person variance or standard deviation is calculated (Carstensen et al., 2000; Krone et al., 2017; Kuppens & Verduyn, 2017). Moreover, it is noteworthy that researchers must consider the duality of emotional changes when they plan to examine emotion variability: the changes of an overall emotional state reflected by the changes in a single variable of emotional intensity, *and* the changes among multiple emotional states (See Fig. 2.1 for an illustration). For instance, as shown in the top right of the figure, one had multiple emotions, but happiness was fairly stable with a couple of sad episodes (i.e., low variability). However, the high variability scenario demonstrated a variety of emotions experienced over time. The research on the variability of multiple emotion categories is still in its infancy but new knowledge is emerging. As a representative example, emotional variability has been defined as the fluctuations in emotional states over time, which can be quantified by entropy analysis (Li et al., 2021a, b).

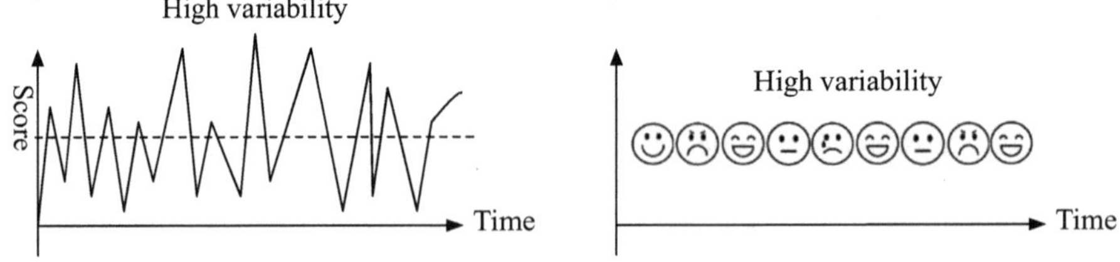

Fig. 2.1 An illustration of emotional variability for a single emotional variable (left) and multiple emotions (right)

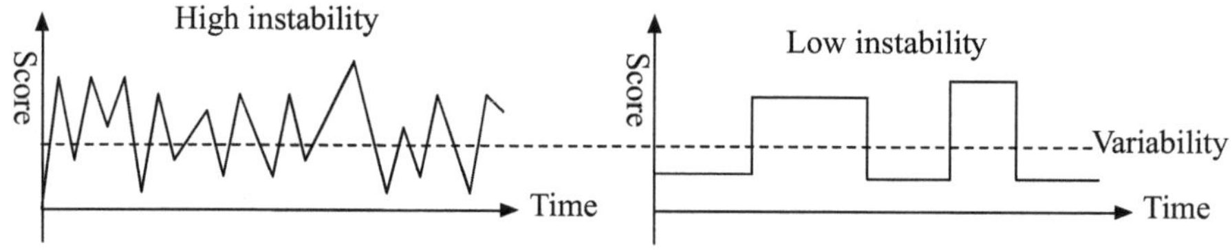

Fig. 2.2 An illustration of the difference between emotional variability and instability

2.2 Emotional Instability

As defined by Houben et al. (2015), emotional instability refers to "the magnitude of emotional changes from one moment to the next" (p. 902). Emotional instability is very similar to emotional variability since both describe the fluctuation of an individual's emotions. Some researchers did not differentiate the two terms. For example, Bailen et al. (2019) used emotional instability and emotional variability interchangeably in a review of emotion in adolescents. However, as shown in Fig. 2.2, the emotional instability of two individuals can be quite different even if they experience the same level of emotional variability. Emotional variability describes the general dispersion of emotional intensity over an entire period, whereas emotional instability captures moment-to-moment changes in emotional intensity. Mathematically, emotional instability is usually calculated as the mean square of successive difference (MSSD) between consecutive emotion scores, the root mean squared successive difference scores (RMSSDs), or the mean absolute successive difference scores (MASDs) (Houben et al., 2015; Reitsema et al., 2022; Sperry et al., 2020).

2.3 Emotional Inertia

Emotional inertia reflects the degree to which an individual's emotional states are resistant to change (Houben et al., 2015; Kuppens et al., 2010; Kuppens & Verduyn, 2017; Reitsema et al., 2022). High emotional inertia means that an individual's emotional state is likely to persist from one moment to the next, and thus is highly predictable. In contrast, low emotional inertia means that an individual's emotional state is more prone to change, suggesting that it is more susceptible to internal or external influences (Kuppens et al., 2010). Emotional inertia is typically operationally defined as the extent to which one's current emotional intensity can be predicted by that of a previous moment (Houben et al., 2015; Kuppens et al., 2010). Consequently, emotional inertia is often calculated as the autocorrelation or autoregressive coefficient of emotions across time (Reitsema et al., 2022).

2.4 Emotional Cross-lags

Emotional cross-lags refer to an important feature of emotion dynamics that is usually operationalized as how the intensity of an emotion influences the intensity of subsequent emotions. Emotional cross-lags occur in two forms: emotional augmentation and emotional blunting (Kuppens & Verduyn, 2015; Reitsema et al., 2022). For the phenomenon of emotional augmentation, the experience of a certain emotion increases the occurrence of another emotion at the next moment. Emotional blunting refers to the phenomenon when a specific emotion blunts or decreases the experience of subsequent emotion(s). As an empirical illustration, Bringmann et al. (2016) found that emotions of the same valence (e.g., relaxed and happy) tended to augment each other, whereas the emotions of different valences blunted each other. Emotional cross-lags are operationalized as the time-lagged cross-correlations or cross-regressive effects between different emotions (Kuppens & Verduyn, 2015; Reitsema et al., 2022). Moreover, network analysis is gaining popularity in assessing emotional augmentation and blunting. For instance, Bringmann et al. (2016) assessed emotional cross-lags and their relation to neuroticism from a network perspective. Specifically, they used a multilevel VAR (vector autoregressive) model to determine the temporal connections among different emotion categories, which were then visualized graphically as an emotion network. A network approach allows researchers to visually pinpoint how different emotions augment or blunt each other over time.

2.5 Emotional Patterns

Emotion dynamics is concerned with the study of "the trajectories, patterns, and regularities with which emotions, or one or more of their subcomponents (such as experiential, physiological, or behavioral components), fluctuate over time, their underlying processes, and downstream consequences" (Kuppens & Verduyn, 2015, p. 72). Defining emotional dynamics in such a broad sense, as we have observed in the field of educational psychology, reflects the interests of educational researchers. However, the most-studied existing features of emotion dynamics, i.e., emotional variability, instability, inertia, and cross-lag, cannot fully capture the patterns and regularities of emotional changes. Thankfully technological and methodological advances have assisted educational researchers in the discovery of emotional patterns. Typical methodological examples include the sequential patterns of emotions revealed by various sequential mining techniques and the recurrence patterns of emotion categories in a time series (Jenkins et al., 2020).

3 The Measurements of Emotion Dynamics

Emotion dynamics can be quantified by a set of temporal features of emotions. There is no direct way to measure emotion dynamics. The measurements of emotion dynamics rely exclusively on the collection of fine-grained emotion data. Therefore, this section describes several prevalent measurement techniques that unobtrusively collect emotion data at a fine-grained size.

3.1 Experience Sampling Method

The experience sampling method (ESM) is an instrument to capture participants' feelings, thoughts, emotions, and actions in the moment with repeated administrations of self-report questionnaires (Zirkel et al., 2015). The ESM can be implemented in three distinct forms: interval-contingent sampling, event-contingent sampling, and signal-contingent sampling (Napa Scollon et al., 2009). The interval-, event-, and signal-contingent samplings occur when participants wait for a designated interval, when they encounter a specific event, and when they are promoted by a randomly timed signal, respectively, to complete self-reports. When using ESM to collect a person's emotion data, researchers can gather dozens or even hundreds of responses regarding the individual's emotional experiences in context. Thus, ESM allows researchers to develop a direct understanding of how and why an individual's emotions change over time within natural settings. The longitudinal and time series data of emotions captured by ESM enable researchers to analyze the patterns of emotional changes. As an example, Sun et al. (2020) asked the participants to complete experience sampling reports of their positive and negative emotions four times per day for 7 days when investigating the fluctuations in emotion experience among 185 participants. However, a shortcoming of ESM is that it can be quite intrusive when participants are consistently prompted to fill in questionnaires.

3.2 Emote-Aloud

The emote-aloud method requires learners to verbalize their affective states in real-time during learning or problem-solving. Prior to the implementation of emote-aloud, participants usually receive training on how to concurrently emote-aloud. Specifically, participants need to focus on their expressions of emotions, and they say out loud whatever emotions they experience in learning. The emote-aloud procedure is typically videotaped or audio-recorded, whereby researchers transcribe the emote-aloud protocols, segment the protocols into meaningful units, and code emotions for each unit (Craig et al., 2008; D'Mello et al., 2006). In a study conducted by Muis et al.

(2020), they used an emote-aloud protocol to capture participants' emotions as they occurred in real-time. Muis et al. (2020) contended that the emote-aloud protocol provided an accurate measure of participants' emotions. First, the coders could refer to the context in which an expression of emotion was labeled. When coding participants' emotions from the transcribed protocols of emote-aloud, Muis et al. (2020) took the sentences immediately before and after into account. In addition to the written transcript, the coders listened to each participant's transcript to assess changes in intonation in their voices to increase coding accuracy. A potential drawback of emote-aloud is that participants can experience emotion unconsciously. Moreover, although emote-aloud is less intrusive compared to ESM, it may distract intense effort as a learner engages in cognitively demanding tasks.

3.3 Facial Expressions

Facial expression provides another important approach for measuring moment-to-moment emotions. Ekman (1993) found evidence of universality in facial expressions across cultures and social situations, whether they be spontaneous or deliberately posed. Therefore, Ekman (1993) contended that it was feasible to detect emotions by modeling the movements of face. The facial movements, reflected by the visible appearance changes in facial muscles, can be described by the Facial Action Coding System (FACS) (Ekman & Friesen, 1976). The FACS involves the identification of action units (AUs), which are the fundamental actions of individual muscles or groups of muscles in the facial expression. Examples of AUs include cheek raiser, inner brow raiser, jaw drop, lip suck, and neck tightener. The FACS has developed to become a standard to comprehensively categorize the physical expression of emotions. Researchers can code emotions from the recorded facial videos, based on the FACS manual. However, it is more common for researchers to assess emotions in real-time using automatic facial expression software embedded with the FACS. For instance, Li et al. (2021a, b) recognized students' emotions using FaceReader, which is a facial expression recognition software that can categorize facial expressions into one of the six basic emotions (i.e., happy, sad, angry, surprised, scared, and disgusted) or a neutral state. Recent years have also witnessed the increasing use of the iMotions FEA (Facial Expression Analysis) module to determine the participants' emotions. Specifically, the iMotions FEA module can recognize seven core emotions, including joy/happiness, confusion/anger, fear, disgust, contempt, sadness, and surprise. Facial expression data is practically entirely unobtrusive except for requirements about positioning of the head for data to be reliably captured.

3.4 Vocal Expressions

Vocal expression of emotion is a phenomenon that describes how the acoustic properties of vocalizations relate to emotional experiences (Bachorowski & Owren, 1995; Scherer et al., 2003). Bachorowski and Owren (1995) explored the feasibility of using acoustic properties of speech, which included the fundamental frequency (F_0), jitter, and shimmer of sound wave, to index emotional processes. They found that both positive and negative emotional states were associated with increases in F_0, and individual differences in emotional intensity mediated participants' vocal expressions of emotion. Bachorowski and Owren (1995) argued that future studies on the characterizations of vocal expression of emotion would benefit from including a wider range of acoustic parameters, such as overall speech rate, energy distribution, and voice amplitude. However, as pointed out by Scherer et al. (2003), much of the work in this area has no solid theoretical foundations, and has been empirically investigating how the inductions of stress or specific emotions in the speaker produce changes in voice and speak production, as well as changes in the patterns of acoustic parameters. Scherer et al. (2003) provided an excellent review of the empirical findings regarding the effect of emotions (i.e., arousal/stress, happiness/elation, anger/rage, sadness, fear/panic, and boredom) on various acoustic parameters. However, some researchers directly use vocal parameters for emotion recognition. For instance, Scherer et al. (2003) found that acoustic signal dimensions, such as duration, amplitude, and energy distribution in the frequency spectrum, were mostly indicative of arousal. Moreover, recent work has attempted to train machine learning models for emotion recognition with selected acoustic features (Kuchibhotla et al., 2014). Using vocal expressions to measure emotion is completely unobtrusive. However, one should be aware that there are currently no established guidelines and mature technologies for the detection of emotion from vocal expressions.

3.5 Language and Discourse

Words and language, as pointed out by Tausczik and Pennebaker (2010), are "the very stuff of psychology and communication" (p. 25). The emotion words learners use provide important cues to their emotional states. There is no surprise that researchers have been attempting to capture participants' emotions from their language use and discourse (Muis et al., 2020; Pennebaker et al., 2015; Xing et al., 2019). For instance, Muis et al. (2020) developed a coding scheme to manually code participants' emotions from the transcribed emote-aloud protocols. Particularly, Muis et al. (2020) claimed that 11 types of emotions could be captured in participants' transcripts, which include anger, anxiety, boredom, confusion, curiosity, enjoyment, sadness, frustration, hopefulness, hopelessness, and relief. However, manual coding is labor-intensive, time-consuming, and potentially unreliable.

Computerized programs that take advantage of natural language processing techniques provide new options for detecting emotions in an automated fashion. As an illustration, the text mining program of the Linguistic Inquiry and Word Count (LIWC) (Pennebaker et al., 2015) provides an efficient method for analyzing learners' emotions, as the program automatically recognizes emotion-related words from the participants' verbal language or writing outputs. Specifically, the LIWC program quantifies positive and negative emotions as the percentages of positive and negative words within a text, respectively. A more recent development of emotion detection in language use and discourse is the use of machine learning algorithms. As an example, Xing et al. (2019) automatically detected the four types of achievement emotions (i.e., positive activating, positive deactivating, negative activating, and negative deactivating) in MOOC (Massive Open Online Courses) forum posts, using supervised machine learning models. Specifically, three kinds of textual features, including language summary features, linguistic features, and Latent Dirichlet Allocation topic features (Blei et al., 2003), were extracted from the forum posts. Taking the three types of textual features for each post as the inputs, and manually coded emotional states as the ground truth, Xing et al. (2019) trained four classic machine learning models (i.e., Naïve Bayes, Logistic Regression, Support Vector Machines, and Decision Tree) for emotion detection. While language and discourse data can be collected in an unobtrusive manner, analyzing such data to extract emotions is methodologically challenging. Moreover, there are great variations among student populations in their use of language and discourse, making it hard to generalize the findings of a study to other contexts.

3.6 Physiological Sensors

Physiological sensors are becoming popular among researchers for making inferences of participants' emotional states in real-time (Harley, 2016). The rationale is that physiological signals reflect the activity of the autonomic nervous system, which is influenced by emotional stimuli (Kim et al., 2004). As a practical example, Kim et al. (2004) reported a physiological signal-based emotion recognition system, which used the signals of electrocardiogram (ECG), skin temperature variation, and electrodermal activity (EDA) to predict concrete emotion categories. As claimed by Kim et al. (2004), the system was developed based on a bio-signal database where the external stimuli, the induced emotional status, and corresponding physiological signals were explicitly labeled. Kim et al. (2004) first extracted emotion-specific characteristics from short-segment signals, based on which they trained a support vector machine to classify emotions. Notably, Koelstra et al. (2011) recorded more physiological signals for emotion analysis, which included the electroencephalogram (EEG) and peripheral nervous system signals, i.e., galvanic skin response (GSR), respiration amplitude, skin temperature, ECG, blood volume by plethysmograph, electromyograms of Zygomaticus and Trapezius muscles, and

Table 2.1 Illustration of different methods to measure emotion dynamics features

	Variability	Instability	Inertia	Cross-lags	Patterns
ESM					
Emote aloud					
Facial expressions					
Vocal expressions					
Language and discourse					
Physiological sensor					

Note: **Light gray: the method can be used to measure emotion dynamics features (EDFs). Dark gray: the method is not ideally suitable for measuring EDFs**
The pattern style of diagonal lines: the method may be appropriate to measure EDFs, but it depends on the devices and the techniques implemented

electrooculogram (EOG). Instead of focusing on the prediction of emotion categories, Koelstra et al. (2011) analyzed the correlations between physiological signals and emotional features. Koelstra et al. (2011) found that EEG scores were powerful indicators of emotional arousal, whereas peripheral nervous system signals were best for the prediction of emotional valence. In short, using physiological sensors to measure emotion allows researchers to capture emotion-related variables continuously at a fine-grained size. Another advantage of physiological sensors lies in their unobtrusiveness. However, researchers need to gain specialized skills to use physiological sensors.

As shown in Table 2.1, we provide an illustration of how those methods can be used to measure different features of emotion dynamics, to assist researchers in making wise decisions.

4 The Techniques for Analyzing Emotion Dynamics

In this section, we provide an overview of the most prominent techniques for analyzing emotion dynamics, including the conventional statistical methods (e.g., variance), entropy analysis, growth curve modeling, time series analysis, network analysis, recurrence quantification analysis, and sequential pattern mining. We acknowledge that the techniques listed above are by no means exhaustive, considering that the field of emotion dynamics is still in its infancy and new analytical techniques are emerging. The aim of this section is to help readers better understand the research base of emotion dynamics and assist researchers to make better analytical decisions by enriching their repertoire of methods and techniques for analyzing different aspects of emotion dynamics.

4.1 Conventional Statistical Methods

The features of emotion dynamics can be mostly expressed through conventional statistical methods. Table 2.2 provides a list of typical methods and techniques for analyzing emotion dynamics features, which includes both conventional statistical methods (e.g., standard deviation and autocorrelation) and advanced analytical techniques. In terms of the conventional statistical methods, the most used metric for assessing emotion variability is standard deviation (*SD*) (Jenkins et al., 2020; Röcke et al., 2009). *SD* is a statistical measure of the amount of variation of a set of values that, in terms of emotions, reflects the magnitude of the change of an individual's emotion scores in relation to the mean. As we discussed before, the assessment of emotional instability is generally based on the scores of either MSSD, RMSSD, or MASD. MSSD is calculated as the mean of the squared difference between successive observations (i.e., consecutive scores of emotional intensity), whereas MASD refers to the average of the absolute difference between successive observations. For emotional inertia, the statistical method of either autocorrelation or autoregressive coefficient has been used for operationalization (Kuppens & Verduyn, 2015; Reitsema et al., 2022). Autocorrelation is the correlation of a time series with its lagged counterpart, and autoregressive coefficient describes the predictive power of past period values to current ones. Similarly, the statistical methods of cross-correlations and cross-regressive effects are usually used to estimate emotional cross-lags.

Table 2.2 Typical methods and techniques for analyzing emotion dynamics features

Features	Variable	Analysis
Emotional variability	Single	Standard deviation Variance
	Multiple	Entropy analysis
Emotional instability	Single	Mean square of successive difference (MSSD) Root mean squared successive difference (RMSSD) Mean absolute successive difference (MASD)
Emotional inertia	Single	Autocorrelation Autoregressive coefficient
Emotional cross-lags	Multiple	Cross-correlations Cross-regressive effects Network analysis
Emotional patterns	Single or multiple	Growth curve modeling Time series analysis
	Multiple	Sequential mining Recurrence quantification analysis

4.2 *Entropy Analysis*

Emotion dynamics can be analyzed from the point of view of information entropy. The concept of information entropy or Shannon entropy (Shannon, 1948) originated from the field of communication but has been applied to study a number of issues, such as diversity in systems (Rajaram et al., 2017), dynamical stability (Cincotta et al., 2021), and the signaling dynamics of facial expressions of emotion (Jack et al., 2014). Li et al. (2021a, b) adopted the Shannon entropy to quantify the randomness of emotional states in a certain time period, which was conceptualized as another indicator of emotion variability, or more directly, emotion entropy. The statistical formula of emotion entropy is the same with the Shannon entropy, as shown below:

$$H\left(p_1,\ldots,p_a\right) = -\sum_{j=1}^{a} p_j \log_2\left(p_j\right)$$

The p_j refers to the probability of an emotional state j (e.g., happy) appearing in a set of emotional states. The minimum value for the emotion entropy is zero, indicating that an individual's emotion never changes. The higher the emotion entropy value, the more variable an individual's emotional states. In sum, entropy analysis of emotion dynamics provides researchers with a straightforward and promising methodological approach. Therefore, we anticipate a gradual but substantial increase in using this analytical method to study emotion dynamics.

4.3 *Growth Curve Modeling*

Growth curve model typically refers to statistical methods that allow for the estimation of inter-individual variability in intra-individual patterns of change over time (Curran et al., 2010). In the context of emotion dynamics, growth curve models can be used to estimate between-person differences in the patterns of emotional changes within each person. We use the study of Ahmed et al. (2013) as an example to illustrate its use in assessing emotion dynamics. In particular, Ahmed et al. (2013) investigated the developmental trends of four academic emotions (i.e., anxiety, boredom, enjoyment, and pride) among 495 students in Grade 7 over a school year, using growth curve analysis. Specifically, Ahmed et al. (2013) used a two-level multilevel modeling technique to estimate the growth trajectories of the four emotions (within-student Level-1 model) and the individual variability in the emotions (between-student Level-2 model). Growth curve analyses revealed that the academic emotions of enjoyment and pride declined, whereas boredom increased over time. Moreover, Ahmed et al. (2013) found meaningful individual variability in the initial levels of both enjoyment and pride.

4.4 Time Series Analysis

Time series analysis is a family of methods that can extract statistically meaningful characteristics from time series data. Time series analysis methods are important alternatives for analyzing emotion dynamics, given that emotions are constantly fluctuating over time and thus are time-dependent and non-stationary. Krone et al. (2017) proposed a vector autoregressive Bayesian dynamic model (VAR-BDM), which can be applied to both univariate and multivariate time series. In this regard, BDM could provide "insights into the dynamics of single emotions as well as the dynamics between multiple emotions within an individual" (Krone et al., 2017, p. 740). More specifically, BDM includes six parameters that are immediate translations of the six features of emotion dynamics, i.e., within-person variability, innovation variability (instability), inertia, cross-lag, granularity, and intensity. Therefore, the analysis results generated by BDM provide a complete picture of emotion dynamics. It is worth mentioning that a fundamental BDM is the VAR (1)-BDM model, which is for stationary individual time series with about normally distributed fluctuations. However, Krone et al. (2017) provided solutions on how to extend the VAR (1)-BDM to deal with non-stationary time series data.

4.5 Network Analysis

Temporal emotion dynamics can be visualized as an emotion network, which consists of nodes (i.e., discrete emotions) and edges that connect the nodes (Bringmann et al., 2016). The thickness of an edge usually indicates the strength of the relationship between the nodes, although the models for inferring the edges vary. A positive or negative value for the edge in an emotion network typically suggests whether the connection between two emotions is positive or negative, respectively. Bringmann et al. (2016) contended that using a network approach for assessing temporal emotion dynamics represents a paradigm shift in our understanding of psychological constructs. Psychological phenomena cannot be fully explained by causal structures of several predefined components. Rather, psychological phenomena are complex systems of interacting components, where the role and strength of relationships between components change over time in nonlinear ways (Hilpert & Marchand, 2018). When it comes to our emotions, they can be conceptualized as a complex dynamical system. The discrete emotions, such as happy, stressed, angry, and sad, interact with each other over time to yield a novel behavioral outcome. As an illustration, Bringmann et al. (2016) found that emotions of the same valence (i.e., positive or negative) tended to augment each other, whereas emotions of different valence seemed to decrease each other. They also found that the temporal interactions of emotions were correlated with neuroticism.

We will not delve into the implementation details of building an emotion network, since Bringmann et al. (2016) have already explained how to analyze emotion dynamics using networks. Researchers who plan the analyses of their own data may want to refer to Bringmann's et al. (2016) work to find the tools, codes, and two demonstrated examples. It is noteworthy that standard autoregressive models can be used for analyses if the statistical hypothesis is met, i.e., the repeatedly measured emotion variables are time-invariant. Otherwise, time-varying autoregressive (TV-AR) models will need to be considered (Bringmann et al., 2017). Moreover, Bringmann et al. (2013) developed a multilevel approach to vector autoregressive (VAR) modeling to extract network structures from nested longitudinal data. The multilevel-VAR model allows for the modeling of emotion dynamics not only within an individual, but also at group level (Bringmann et al., 2013, 2016).

4.6 Recurrence Quantification Analysis

The last few years have seen the introduction of recurrence quantification analysis (RQA) in educational research (Fleuchaus et al., 2020; Li et al., 2022; Wallot, 2017). As pointed out by Fleuchaus et al. (2020), dynamic stability is "a well-defined construct that can be indexed precisely…(and) RQA can determine the presence of dynamic stabilities by analyzing variability in time-series data" (p. 448). Specifically, RQA is a non-linear analysis that can assess the repetition of elements in a time series with a range of metrics, such as percent recurrence (*%REC*) and percent determinism (*%DET*). RQA metrics are calculated based on recurrence plot. Recurrence plot is a visualization of the recurrence values within a discrete time series by plotting the time series on both the x- and y-axis of a two-dimensional grid. Figure 2.3 shows the illustration of a recurrence plot. When applying RQA on a time series of emotion states, *%REC* measures the degree to which the same state of emotion reoccurs over time. For instance, the happy emotion may reoccur 50 times within an affect data series, and the hopeless emotion may reoccur 30 times. *%REC* is calculated by dividing the total recurrence time by $N(N\text{-}1)$, where N refers to the length of a time series. *%DET* is a measure of regularity that reflects the degree to which the same (or similar) sequences of affective change over time (Jenkins et al., 2020). Examples of the same sequences of affective change include "happy-curious-disappointed-boredom", "surprised-boredom-hopeless", and "anxiety-happy-excited-relief". Therefore, *%DET* represents the degree of affect predictability or deterministic structure within a time series of emotions (Jenkins et al., 2020). Researchers who are interested in RQA may find Wallot's (2017) work helpful, where he provided a step-by-step tutorial on how to run RQA using R, as well as some guidance regarding common issues and best practices using RQA.

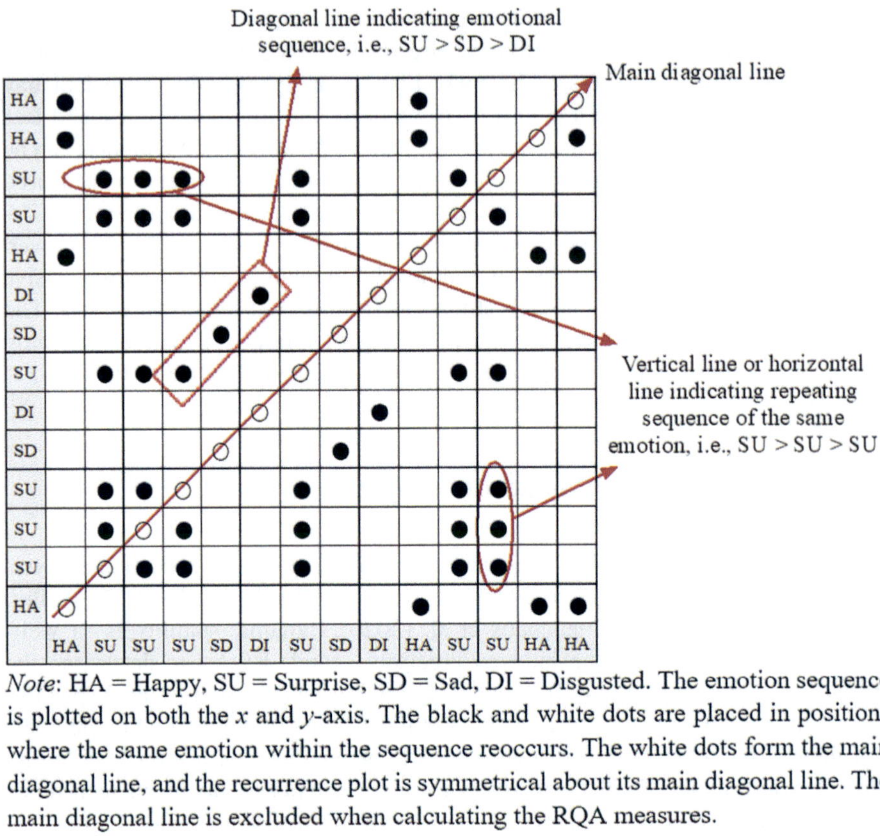

Fig. 2.3 An illustration of recurrence plot
Note: *HA* Happy, *SU* Surprise, *SD* Sad, *DI* Disgusted. The emotion sequence is plotted on both the x- and y-axis. The black and white dots are placed in positions where the same emotion within the sequence reoccurs. The white dots form the main diagonal line, and the recurrence plot is symmetrical about its main diagonal line. The main diagonal line is excluded when calculating the RQA measures

4.7 Sequential Pattern Mining

Sequential pattern mining techniques are gaining popularity to analyze emotion sequences that consist of multiple emotional categories in time order. In general, the process of sequential pattern mining is to extract the frequently occurring patterns in a sequence that exceeds a predefined minimal support threshold. Sequential pattern mining can reveal the relationships between occurrences of emotions in a time series, and whether there exist any specific orders of the occurrences. As an example, Lajoie et al. (2019) used the lag sequential analysis to examine the patterns of participants' emotion sequences as they solved clinical reasoning problems with an intelligent tutoring system. Furthermore, Lajoie et al. (2019) visualized

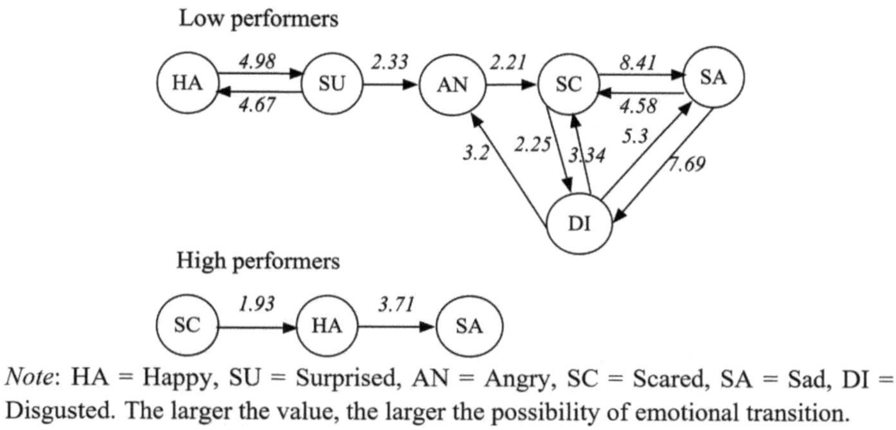

Fig. 2.4 The patterns of emotion sequences of low performers (top) and high performers (bottom) *Note*: *HA* Happy, *SU* Surprised, *AN* Angry, *SC* Scared, *SA* Sad, *DI* Disgusted. The larger the value, the larger the possibility of emotional transition

participants' emotion transition patterns as diagrams for easy interpretation of group differences (see Fig. 2.4). They found that happiness was followed by anger, scared, disgust, and sadness for low performers. In addition, the emotion transition patterns of low performers were more variable and unpredictable than that of high performers.

5 The Challenges of Studying Emotion Dynamics in Learning

The research on emotion dynamics in education is inevitably influenced by the contemporary literature, which has a deep root in psychopathology and psychological well-being. Although the benefits are many, several barriers are expected for educational researchers on the way to researching emotion dynamics by leveraging the existing research. For one, educational researchers may need to re-examine and redefine the core concepts related to emotion dynamics since they are not necessarily domain general. For example, a number of studies defined emotional instability, often used interchangeably with affective instability, as a type of emotional dysregulation (Marwaha et al., 2014). Researchers encounter a dilemma regarding whether to continue using the existing constructs related to emotion dynamics (e.g., emotional variability, instability, and inertia) or create a new taxonomy of emotion dynamics for educational research. For the purpose of this chapter, we highlight some key challenges associated with the measurement and analysis of emotion dynamics in learning.

5.1 Deciding What to Measure About Emotion Dynamics

According to Houben et al. (2015), the variability, instability, and inertia of emotions are the three most studied attributes of emotion dynamics. However, several questions naturally arise when measuring these attributes in teaching and learning contexts: Do the three emotion dynamics attributes provide a complete picture of an individual's emotional changes? Are there any features of emotion dynamics that are crucial to students' learning? Is emotional inertia a good measure of emotion dynamics in a learning activity? and so forth.

In a meta-analytic and descriptive review of emotion dynamics in children and adolescents, Reitsema et al. (2022) provided a table of emotion dynamics measures, including intensity, variability, instability, inertia, differentiation or granularity, and augmentation and blunting. While the work of Reitsema et al. (2022) provides new insights about emotion dynamics patterns, we contend that the features of emotion dynamics should be differentiated from an individual's ability to recognize and regulate their emotions. Particularly, emotion differentiation or emotion granularity refers to an individual's ability to make nuanced distinctions between similar emotional states (Smidt & Suvak, 2015). Based on the definition of emotion differentiation, it does not necessarily reflect the changes of emotions. Nevertheless, emotion differentiation is "often operationalized as emotional covariance or dependencies and co-occurrences between multiple emotions" (Reitsema et al., 2022, p. 377). The operational definition of emotion differentiation, however, describes how emotions interact with each other over time. In this regard, emotion differentiation can be considered as a feature of emotion dynamics. All in all, in addition to the conceptual ambiguities between the features of emotion dynamics and individuals' emotional capacity, the mismatch of the conceptual and operational definitions of emotion dynamics features calls for more attention and studies in this area.

5.2 Deciding How to Analyze Emotion Dynamics

The heterogeneous methodologies for analyzing emotion dynamics present many decision-making challenges for educational researchers, especially for those who do not have a clear understanding of currently available analytical techniques. Researchers choose different analytical techniques based on the nature of the phenomenon, the research questions, the data available, and their preferences and skill sets. Therefore, the operational definition of an attribute of emotion dynamics can vary significantly across studies. For example, emotional variability can be quantified as either the variance or standard deviation of an individual's emotional intensity across time. It is also helpful to analyze emotional variability with entropy analysis (Li et al., 2021a). Consequently, researchers will need to make themselves aware of the insights that can be obtained from their chosen technique, as well as its shortcomings. Moreover, advanced techniques for analyzing emotion dynamics are emerging, adding another level of challenge to researchers' decision-making process.

5.3 *Addressing Individual and Developmental Differences*

The generalizability of the study findings about emotion dynamics affects the adoption and recognition of this area of research among researchers, learners, and practitioners. Adding to the challenge is the fact that there are significant individual and developmental differences in emotion dynamics. Learners differ in how they appraise their learning and problem-solving circumstances. There are also individual differences in the emotion-appraisal system, which establishes how appraisals or patterns of appraisal components relate to specific emotional experiences (Kuppens et al., 2009). Additionally, the emotion-appraisal relationships change through the lifespan, given that the emotion appraisal and emotion regulation skills of an individual tend to mature and develop over time. For instance, Reitsema et al. (2022) revealed systematic changes in emotion dynamics throughout childhood and adolescence, in a meta-analytic and descriptive review of 102 ecological momentary assessment studies that involved 19,928 participants. As an illustration, Reitsema et al. (2022) found the instability of both positive and negative emotions decreases from early to late adolescence. Therefore, it is crucial to carefully consider the selection of computational models that can account for individual and development differences when analyzing emotion dynamics.

5.4 *Differentiating Between Short-Term and Long-Term Emotion Dynamics*

The boundary between short-term and long-term emotion dynamics blurs, which presents another challenge to investigate the dynamical nature of emotions. As pointed out by Houben et al. (2015), emotion dynamics can be examined on varying time scales. However, the techniques and instruments for studying emotional changes across seconds or minutes are undoubtedly different from those for exploring affective changes over days or several years. Therefore, researchers should clearly define the period for which students' emotion dynamics will be studied. Researchers are also expected to develop an understanding of whether the features of emotion dynamics examined in their studies reflect more state-like or trait-like individual differences. It is also worth mentioning that learners demonstrate differences in emotional flexibility, which refers to an individual's ability to respond flexibly to changing circumstances (Kashdan & Rottenberg, 2010). In this regard, researchers may find themselves unable to differentiate between short-term (state-like) and long-term (trait-like) emotion dynamics based on simple time scales. Emotional flexibility is another factor that can obscure state-like and trait-like emotion dynamics.

6 Concluding Remarks and Directions for Future Research

Educational researchers are only beginning to examine emotion dynamics. We present a taxonomy of emotion dynamics features, to help educational researchers rethink these features as a first step in considering educational interventions. We then provided a review of measurements and techniques for studying emotion dynamics, which could potentially advance this field of study from a practical standpoint. Considering the lack of theoretical groundwork for this type of research and a shortage of empirical studies on emotion dynamics, there are challenges for connecting this work in the context of teaching and learning. However, those challenges present new opportunities for the development of theoretical frameworks, models, and approaches that can support the design of scaffolding and interventions related to emotion dynamics features. Specifically, an important direction for future research is to develop a better theoretical framework that helps explain emotion dynamics. This framework could become an interdisciplinary effort that joins personality psychology, developmental science, learning science, and neuroscience. Another direction for future research is to examine how emotions fluctuate across different learning phases. It is crucial to unravel the mechanisms of emotion dynamics in various learning processes for the design of effective scaffolding and intervention strategies. For example, Li et al. (2021a) examined the joint effect of emotional variability and the frequency of emotions at each phase of self-regulated learning (i.e., forethought, performance, and self-reflection) on students' clinical reasoning performance. They found that emotional variability negatively predicted performance regardless of which SRL (self-regulated learning) phase it was tied to. Future studies tying emotion dynamics to SRL are needed. Furthermore, empirical investigations of how emotion dynamics features are attached to learning activities, such as goal setting, self-observation, causal attribution, and strategic adaptation are needed.

References

Ahmed, W., van der Werf, G., Kuyper, H., & Minnaert, A. (2013). Emotions, self-regulated learning, and achievement in mathematics: A growth curve analysis. *Journal of Educational Psychology, 105*(1), 150–161. https://doi.org/10.1037/a0030160

Bachorowski, J.-A., & Owren, M. J. (1995). Vocal expression of emotion: Acoustic properties of speech are associated with emotional intensity and context. *Psychological Science, 6*(4), 219–224.

Bailen, N. H., Green, L. M., & Thompson, R. J. (2019). Understanding emotion in adolescents: A review of emotional frequency, intensity, instability, and clarity. *Emotion Review, 11*(1), 63–73.

Blei, D. M., Ng, A. Y., & Jordan, M. I. (2003). Latent dirichlet allocation. *Journal of Machine Learning Research, 3*, 993–1022.

Bringmann, L. F., Vissers, N., Wichers, M., Geschwind, N., Kuppens, P., Peeters, F., Borsboom, D., & Tuerlinckx, F. (2013). A network approach to psychopathology: New insights into clinical longitudinal data. *PLoS One, 8*(4), e60188.

Bringmann, L. F., Pe, M. L., Vissers, N., Ceulemans, E., Borsboom, D., Vanpaemel, W., Tuerlinckx, F., & Kuppens, P. (2016). Assessing temporal emotion dynamics using networks. *Assessment, 23*(4), 425–435.

Bringmann, L. F., Hamaker, E. L., Vigo, D. E., Aubert, A., Borsboom, D., & Tuerlinckx, F. (2017). Changing dynamics: Time-varying autoregressive models using generalized additive modeling. *Psychological Methods, 22*(3), 409.

Carstensen, L. L., Pasupathi, M., Mayr, U., & Nesselroade, J. R. (2000). Emotional experience in everyday life across the adult life span. *Journal of Personality and Social Psychology, 79*(4), 644.

Cincotta, P. M., Giordano, C. M., Silva, R. A., & Beaugé, C. (2021). The Shannon entropy: An efficient indicator of dynamical stability. *Physica D: Nonlinear Phenomena, 417*, 132816.

Craig, S. D., D'Mello, S., Witherspoon, A., & Graesser, A. (2008). Emote aloud during learning with AutoTutor: Applying the facial action coding system to cognitive–affective states during learning. *Cognition and Emotion, 22*(5), 777–788.

Curran, P. J., Obeidat, K., & Losardo, D. (2010). Twelve frequently asked questions about growth curve modeling. *Journal of Cognition and Development, 11*(2), 121–136. https://doi.org/10.1080/15248371003699969

D'Mello, S. K., Craig, S. D., Sullins, J., & Graesser, A. C. (2006). Predicting affective states expressed through an emote-aloud procedure from AutoTutor's mixed-initiative dialogue. *International Journal of Artificial Intelligence in Education, 16*(1), 3–28.

Ekman, P. (1993). Facial expression and emotion. *American Psychologist, 48*(4), 384.

Ekman, P., & Friesen, W. V. (1976). Measuring facial movement. *Environmental Psychology and Nonverbal Behavior, 1*(1), 56–75.

Fleuchaus, E., Kloos, H., Kiefer, A. W., & Silva, P. L. (2020). Complexity in science learning: Measuring the underlying dynamics of persistent mistakes. *Journal of Experimental Education, 88*(3), 448–469. https://doi.org/10.1080/00220973.2019.1660603

Gross, J. J. (2013). Emotion regulation: Conceptual and empirical foundations. In J. J. Gross (Ed.), *Handbook of emotion regulation (2nd ed., pp. 3–20)*. Guilford Publications.

Harley, J. M. (2016). Measuring emotions: A survey of cutting edge methodologies used in computer-based learning environment research. In S. Y. Tettegah & M. Gartmeier (Eds.), *Emotions, technology, design, and learning (pp. 89–114)*. Academic Press. https://doi.org/10.1016/B978-0-12-801856-9.00005-0

Hilpert, J. C., & Marchand, G. C. (2018). Complex systems research in educational psychology: Aligning theory and method. *Educational Psychologist, 53*(3), 185–202. https://doi.org/10.1080/00461520.2018.1469411

Houben, M., Van Den Noortgate, W., & Kuppens, P. (2015). The relation between short-term emotion dynamics and psychological Well-being: A meta-analysis. *Psychological Bulletin, 141*(4), 901.

Jack, R. E., Garrod, O. G. B., & Schyns, P. G. (2014). Dynamic facial expressions of emotion transmit an evolving hierarchy of signals over time. *Current Biology, 24*(2), 187–192.

Jenkins, B. N., Hunter, J. F., Richardson, M. J., Conner, T. S., & Pressman, S. D. (2020). Affect variability and predictability: Using recurrence quantification analysis to better understand how the dynamics of affect relate to health. *Emotion, 20*(3), 391–402. https://doi.org/10.1037/emo0000556

Kashdan, T. B., & Rottenberg, J. (2010). Psychological flexibility as a fundamental aspect of health. *Clinical Psychology Review, 30*(7), 865–878.

Kim, K. H., Bang, S. W., & Kim, S. R. (2004). Emotion recognition system using short-term monitoring of physiological signals. *Medical and Biological Engineering and Computing, 42*(3), 419–427.

Koelstra, S., Muhl, C., Soleymani, M., Lee, J.-S., Yazdani, A., Ebrahimi, T., Pun, T., Nijholt, A., & Patras, I. (2011). Deap: A database for emotion analysis using physiological signals. *IEEE Transactions on Affective Computing, 3*(1), 18–31.

Krone, T., Albers, C. J., Kuppens, P., & Timmerman, M. E. (2017). A multivariate statistical model for emotion dynamics. *Emotion, 18*(5), 739–754. https://doi.org/10.1037/emo0000384

Kuchibhotla, S., Vankayalapati, H. D., Vaddi, R. S., & Anne, K. R. (2014). A comparative analysis of classifiers in emotion recognition through acoustic features. *International Journal of Speech Technology, 17*(4), 401–408.

Kuppens, P., & Verduyn, P. (2015). Looking at emotion regulation through the window of emotion dynamics. *Psychological Inquiry, 26*(1), 72–79.

Kuppens, P., & Verduyn, P. (2017). Emotion dynamics. *Current Opinion in Psychology, 17*, 22–26. https://doi.org/10.1016/j.copsyc.2017.06.004

Kuppens, P., Stouten, J., & Mesquita, B. (2009). Individual differences in emotion components and dynamics: Introduction to the special issue. *Cognition and Emotion, 23*(7), 1249–1258.

Kuppens, P., Allen, N. B., & Sheeber, L. B. (2010). Emotional inertia and psychological maladjustment. *Psychological Science, 21*(7), 984–991.

Lajoie, S. P., Zheng, J., Li, S., Jarrell, A., & Gube, M. (2019). Examining the interplay of affect and self regulation in the context of clinical reasoning. *Learning and Instruction, 101219*, 101219. https://doi.org/10.1016/j.learninstruc.2019.101219

Li, S., Zheng, J., & Lajoie, S. P. (2021a). The frequency of emotions and emotion variability in self-regulated learning: What matters to task performance ? *Frontline Learning Research, 9*(4), 76–91.

Li, S., Zheng, J., Lajoie, S. P., & Wiseman, J. (2021b). Examining the relationship between emotion variability, self-regulated learning, and task performance in an intelligent tutoring system. *Educational Technology Research and Development*, 1–20. https://doi.org/10.1007/s11423-021-09980-9

Li, S., Zheng, J., Huang, X., & Xie, C. (2022). Self-regulated learning as a complex dynamical system: Examining students' STEM learning in a simulation environment. *Learning and Individual Differences, 95*, 102144. https://doi.org/10.1016/j.lindif.2022.102144

Marwaha, S., He, Z., Broome, M., Singh, S. P., Scott, J., Eyden, J., & Wolke, D. (2014). How is affective instability defined and measured? A systematic review. *Psychological Medicine, 44*(9), 1793–1808.

Muis, K. R., Etoubashi, N., & Denton, C. A. (2020). The catcher in the lie: The role of emotions and epistemic judgments in changing students' misconceptions and attitudes in a post-truth era. *Contemporary Educational Psychology, 62*, 101898.

Napa Scollon, C., Prieto, C.-K., & Diener, E. (2009). Experience sampling: Promises and pitfalls, strength and weaknesses. In E. Diener (Ed.), *Assessing Well-being: The collected works of Ed Diener (pp. 157–180)*. Springer.

Oliver, M. N. I., & Simons, J. S. (2004). The affective lability scales: Development of a short-form measure. *Personality and Individual Differences, 37*, 1279–1288. https://doi.org/10.1016/j.paid.2003.12.013

Pekrun, R. (2006). The control-value theory of achievement emotions: Assumptions, corollaries, and implications for educational research and practice. *Educational Psychology Review, 18*(4), 315–341. https://doi.org/10.1007/s10648-006-9029-9

Pekrun, R., Goetz, T., Titz, W., & Perry, R. P. (2002). Academic emotions in students' self-regulated learning and achievement: A program of qualitative and quantitative research. *Educational Psychologist, 37*(2), 91–105.

Pennebaker, J. W., Boyd, R. L., Jordan, K., & Blackburn, K. (2015). *The development and psychometric properties of LIWC2015*. University of Texas at Austin.

Rajaram, R., Castellani, B., & Wilson, A. N. (2017). Advancing Shannon entropy for measuring diversity in systems. *Complexity, 8715605*, 1. https://doi.org/10.1155/2017/8715605

Reitsema, A. M., Jeronimus, B. F., van Dijk, M., & de Jonge, P. (2022). Emotion dynamics in children and adolescents: A meta-analytic and descriptive review. *Emotion, 22*(2), 374–396. https://doi.org/10.1037/emo0000970

Röcke, C., Li, S.-C., & Smith, J. (2009). Intraindividual variability in positive and negative affect over 45 days: Do older adults fluctuate less than young adults? *Psychology and Aging, 24*(4), 863.

Scherer, K. R., Johnstone, T., & Klasmeyer, G. (2003). Vocal expression of emotion. In R. J. Davidson, K. R. Scherer, & H. H. Goldsmith (Eds.), *Handbook of affective sciences (pp. 433–456)*. Oxford University Press.

Schutz, P. A., & Davis, H. A. (2000). Emotions and self-regulation during test taking. *Educational Psychologist, 35*(4), 243–256. https://doi.org/10.1207/S15326985EP3504

Shannon, C. E. (1948). A mathematical theory of communication. *The Bell System Technical Journal, 27*(3), 379–423.

Smidt, K. E., & Suvak, M. K. (2015). A brief, but nuanced, review of emotional granularity and emotion differentiation research. *Current Opinion in Psychology, 3*, 48–51.

Sperry, S. H., Walsh, M. A., & Kwapil, T. R. (2020). Emotion dynamics concurrently and prospectively predict mood psychopathology. *Journal of Affective Disorders, 261*, 67–75.

Sun, J., Schwartz, H. A., Son, Y., Kern, M. L., & Vazire, S. (2020). The language of Well-being: Tracking fluctuations in emotion experience through everyday speech. *Journal of Personality and Social Psychology, 118*(2), 364.

Tausczik, Y. R., & Pennebaker, J. W. (2010). The psychological meaning of words: LIWC and computerized text analysis methods. *Journal of Language and Social Psychology, 29*(1), 24–54. https://doi.org/10.1177/0261927X09351676

Wallot, S. (2017). Recurrence quantification analysis of processes and products of discourse: A tutorial in R. *Discourse Processes, 54*(5–6), 382–405.

Xing, W., Tang, H., & Pei, B. (2019). Beyond positive and negative emotions: Looking into the role of achievement emotions in discussion forums of MOOCs. *The Internet and Higher Education, 43*, 100690.

Zheng, J., Huang, L., Li, S., Lajoie, S. P., Chen, Y., & Hmelo-Silver, C. E. (2021). Self-regulation and emotion matter: A case study of instructor interactions with a learning analytics dashboard. *Computers & Education, 161*, 104061.

Zirkel, S., Garcia, J. A., & Murphy, M. C. (2015). Experience-sampling research methods and their potential for education research. *Educational Researcher, 44*(1), 7–16. https://doi.org/10.3102/0013189X14566879

Chapter 3
Applying Log Data Analytics to Measure Problem Solving in Simulation-Based Learning Environments

Karen D. Wang, Shima Salehi, and Carl Wieman

Abstract Interactive tasks embedded in open-ended digital learning environments offer a promising approach to measuring students' higher-order competencies efficiently and at scale. More research is needed at the intersection of learning analytics and educational measurement to make these interactive tasks useful assessments in classrooms. This chapter represents our research efforts toward understanding how the log data of students' interactions within an educational simulation can be translated into meaningful evidence about their problem-solving process. Our analyses reveal that features extracted from log data are both significant predictors of students' problem-solving outcomes and indicators of specific problem-solving practices. Specifically, instances of deliberate pause during the problem-solving process could be an important and generalizable feature associated with students' problem-solving competencies across different tasks. The results highlight the utility of log data generated in interactive learning environments to provide unobtrusive observations of students' problem-solving processes and the power of learning analytics techniques to extract semantically meaningful behavior patterns associated with specific problem-solving practices.

Keywords Problem-solving · Open-ended learning environments · Simulations · Log data · Learning analytics · Educational data mining

---

K. D. Wang (✉) · S. Salehi
Graduate School of Education, Stanford University, Stanford, CA, USA
e-mail: kdwang@stanford.edu; salehi@stanford.edu

C. Wieman
Graduate School of Education, Stanford University, Stanford, CA, USA

Department of Physics, Stanford University, Stanford, CA, USA
e-mail: cwieman@stanford.edu

V. Kovanovic et al. (eds.), *Unobtrusive Observations of Learning in Digital Environments*, Advances in Analytics for Learning and Teaching,
https://doi.org/10.1007/978-3-031-30992-2_3

1 Introduction

As advances in artificial intelligence take over well-defined, routine tasks, the ability to solve complex, unstructured problems becomes an increasingly important and (so far) uniquely human endeavor (Levy & Murnane, 2013). The US National Research Council has recognized this trend and listed practices related to problem-solving at the core of the Next Generation Science Standards (NGSS Lead States, 2013; Holthuis et al., 2018). The ABET (Accreditation Board for Engineering and Technology) states that learning how to solve complex problems is an essential part of engineering education (ABET, 2022). The Organization for Economic Co-operation and Development (OECD) incorporates items assessing problem-solving competence into the Programme for International Student Assessment (PISA) (OECD, 2014; Csapó & Funke, 2017; Stadler et al., 2020). Despite the growing consensus that teaching problem solving should be a key component of science and engineering education, the development of innovative learning and assessment activities progress slowly. Even the most advanced educational technology solutions today are challenged to reliably and validly measure students' competencies in problem solving.

Problem solving can be broadly defined as the cognitive and metacognitive processes that one goes through to reach a goal when the series of actions in the solution path is not immediately available (Newell & Simon, 1972; OECD, 2014). The specific steps and practices involved in solving a problem are largely dependent on the nature of the problem. To teach students the practices used by scientists and engineers to solve real-world problems, we must first explicate the characteristics of authentic problems in science and engineering domains. These problems bear little resemblance to the exercise questions in textbooks and exams (Price et al., 2022). Instead, they share the following features: (1) *providing insufficient initial data*: authentic problems provided no or only limited data upfront and it is up to the problem solver to decide what data to collect and how to collect the data to better define and solve the problem; (2) *requiring domain knowledge*: solving these problems requires the application of domain-specific knowledge and it is up to the problem solver to decide what concepts/formula/predictive framework to apply; (3) *prescribing no solution path or criteria for success*: these problems do not come with a prescribed path to reach a solution or specify the criteria for evaluating a solution. Problem solvers must decide for themselves what actions to take to reach a solution and what criteria to use to evaluate success (Salehi, 2018). With these characteristics in mind, our research group designed and developed a set of interactive problem-solving tasks embedded in PhET simulations (www.phet.colorado.edu).

In our previous research, we conducted qualitative analyses on the video recordings of experts and students solving one of the interactive problems (see details of the black box problem in the Methods section) to precisely define the specific practices involved in solving such authentic problems in science and engineering domains. A framework of problem-solving practices emerged from the analyses and includes the following elements (Salehi, 2018):

- Problem definition and decomposition: these are practices that problem solvers engage in to understand and simplify a problem, such as articulating a problem in one's own words and breaking down a problem into smaller subproblems that are easier to solve.
- Data collection: this practice refers to the actions and decision-making that problem solvers engage in to collect the data needed to solve a problem.
- Data recording: this practice refers to how problem solvers keep track of the data collected.
- Data interpretation: this practice refers to how problem solvers apply domain knowledge to make sense of the data collected and reach a solution.
- Reflection: this encompasses the cognitive and metacognitive processes that problem solvers engage in to monitor their problem-solving progress and evaluate the quality of their solution, including reflection on problem definition and assumptions, reflection on knowledge, reflection on strategy, and reflection on solution.

A subset of these practices has also been identified by previous research work on scientific inquiry and problem solving (Polya, 1971; OECD, 2005; Wu & Adams, 2006; Windschitl et al., 2008; Pedaste et al., 2015). While the problem-solving practices framework gives us a clear view of what to look for in analyzing students' problem-solving processes, scoring the practices through video recordings of individual students' solution processes is both labor-intensive and subject to human error. Our current project explores how to automate the assessment of these practices through the log files of students' interaction data. The following research questions are addressed in this chapter:

- RQ1. How to extract meaningful behavioral patterns, or features, from the logged interaction data of students solving an open-ended problem in a simulation-based learning environment?
- RQ2. To what extent are the features extracted from log data associated with specific problem-solving practices and general problem-solving outcomes?

2 Background

Interactive tasks embedded in open-ended learning environments (OELEs) offer a promising approach for capturing and measuring students' higher-order competencies. Digital OELEs are integrated systems that provide interactive, learner-centered activities that can engage students in complex, authentic inquiry and problem-solving (Hannafin & Land, 1997; Land & Jonassen, 2012). As one type of OELEs, simulations are interactive computer programs that contain models of scientific phenomena or engineered systems (de Jong & van Joolingen, 1998; Wieman et al., 2008). Educational simulations like PhET Interactive Simulations allow students to explore scientific phenomena and solve problems in an authentic, safe, and cost-effective manner. Key characteristics of such simulations include open-ended interactivity, dynamic and visual display of phenomena, and removal of sources of extraneous cognitive load associated with physical lab equipment. In addition, the interaction data logged in the simulation platform allows for unobtrusive observations of students' work processes. These features make educational simulations like PhET promising platforms for hosting tasks designed to capture the multifaceted practices used to solve authentic problems in science and engineering domains.

In the context of interactive learning environments, log files may contain a time-stamped sequence of student interactions as well as the states and parameter changes of the underlying model. Compared to traditional learning and assessment tasks that only capture the outcome of problem-solving, log data generated by OELE-based tasks provides detailed information on the processes that students go through to solve a problem. Furthermore, log data is automatically collected in a manner that does not interfere with students' natural work process. However, the large volume of unstructured log data does not directly constitute evidence for students' problem-solving competencies. As highlighted in a report by the US National Research Council, "the most important technical challenge to embedding assessment in simulation games is how to make use of the rich stream of data and complex patterns as learners interact with these technologies" (National Research Council, 2011, p. 99).

Analyzing the log data generated in OELEs to extract insights into students' cognitive and metacognitive processes is an active area of research in learning analytics and educational data mining (Fischer et al., 2020; Wang et al., 2023). Despite progress, the research work linking interactive tasks to educational assessments faces several challenges, one of which is the lack of validity, reliability, and generalizability of the inferences made about students' competence based on their performance in these tasks (Gašević et al., 2022). We propose that the following factors may have contributed to this challenge.

First, there exists considerable technical complexity in processing and parsing a vast amount of unstructured log data generated as students work through a task in their own ways. Second, there is a lack of general principles and workflow for identifying semantically meaningful features with assessment and instructional values (NRC, 2011). Kardan & Conati (2011) proposed a framework for identifying meaningful patterns from students' interactions logged in digital learning environments and using the patterns to group students into different profiles. The framework is generalizable to the extent of classifying students based on their general effectiveness in inquiry as measured by knowledge gain (Fratamico et al., 2017), yet cannot predict the effectiveness of specific practices. Third, researchers tend to focus on the overt actions taken by students when working on an OELE-based task, such as clicking on a specific user interface (UI) element. This leads to the features and behavioral patterns extracted from log data being highly specific to the task and OELE used, making it challenging to validate the features and generalize the findings across different tasks and learning environments. Furthermore, as authentic problems do not come with a prescribed solution path that students could mindlessly follow, solving them necessitates the interplay between thinking and doing, exploration and reflection. Focusing on the on-screen actions and overlooking the periods of inactivity during students' work processes would risk missing the opportunity to infer key cognitive and metacognitive processes associated with problem solving.

The current study takes a step toward addressing the above challenges by adopting a theory-driven approach to processing the log data generated in OELE-based problem-solving tasks. We seek to understand how expert knowledge can help parse the log data and extract semantically meaningful features that map to both students' general problem-solving outcomes (i.e., whether they can obtain the correct solution) and their adoption of specific problem-solving practices. Of particular interest to us is to explore pause as a potentially generalizable feature to be extracted from the log data of different tasks. Our goal is not to define pause as a problem-solving practice but to propose guidelines and techniques on how to leverage pause analysis to investigate the cognitive and metacognitive processes of problem solving.

3 Methods

Our research group designed and developed a set of interactive problem-solving tasks embedded in the PhET simulations to mimic authentic problems in science and engineering domains. We present two of these tasks used in two separate experiments below: the black box problem in the Circuit Construction Kit simulation and the mystery gift problem in the Balancing Act simulation.

3.1 Experiment 1

Materials The black box problem embedded in the PhET Circuit Construction Kit (CCK) simulation is an interactive task that preserves the essential characteristics of troubleshooting a circuit. The goal of this problem is to infer the circuit configuration hidden behind a black box by interacting with the four wires ("terminals") protruding from the box (Fig. 3.1). Solving the problem requires knowledge of basic electric circuits and Ohm's law. In addition, the data needed for solving the problem is not provided upfront. Instead, students have to decide what data to collect and how to collect the data through interacting with the simulation. Lastly, the problem does not specify what the solution may look like or the criteria for a correct solution. Students are asked to draw a circuit diagram representing their solutions for the hidden circuit at the end of their problem-solving process. These features make the black box problem resemble a real-world troubleshooting problem more than a typical textbook problem about making calculations using Ohm's law. At the same time, the simulation reduces the complexity associated with real-world troubleshooting by simplifying the electrical components involved in the task, minimizing the chance of measurement errors, and making the invisible information (e.g., electron flow) visible to students through animations.

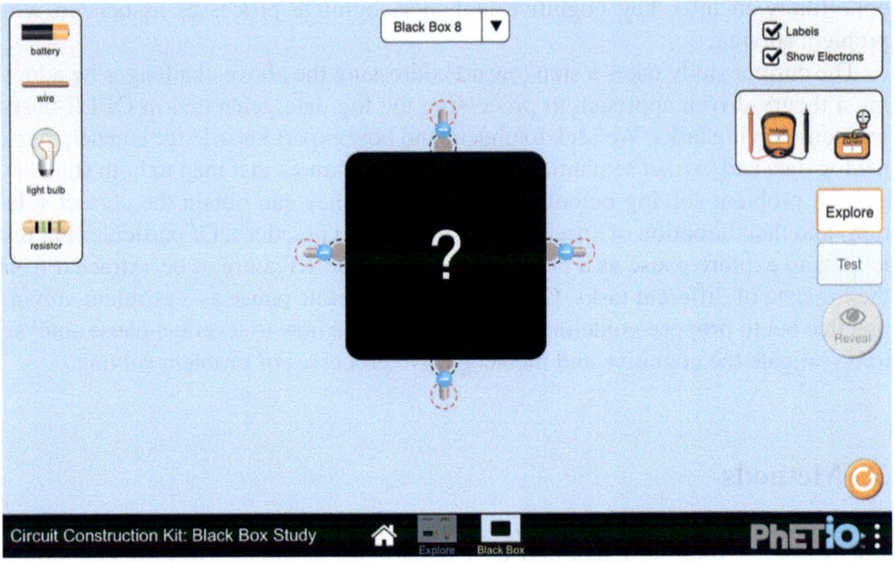

Fig. 3.1 The black box problem asks students to figure out the hidden circuit by building circuits across the terminals and taking measurements. (Image by PhET Interaction Simulations, licensed under CC-BY 4.0)

Participants Seventy-two undergraduate students (58% female) were recruited via email listservs at a highly selective R1 university and participated in the study in an in-person, one-on-one interview setting. To qualify for the study, students must have taken a high-school or college-level physics course covering electricity but not major in physics or electrical engineering. This inclusion criterion ensures that participants have a moderate amount of knowledge in electrical circuits. Around 2/3 of the participants were students in science, technology, engineering, and mathematics (STEM) majors, while the other 1/3 were humanities and social sciences majors.

Procedures Participants worked on the black box problem on a computer and were provided with a calculator, pen, and paper for calculations and notetaking. After informed consent, the researcher gave a brief tutorial to help participants navigate different features of the simulation and refresh their knowledge about Ohm's law by instructing them to build a circuit using different electrical components and take measurements using the ammeter and voltmeter. Participants were then given 15 min to solve the first black box problem and instructed to think out loud while solving it. The researchers interfered minimally during participants' problem-solving process, doing so only to remind them to think aloud or that they were running out of time. Participants drew a diagram of what they thought was hidden behind the black box on paper when they reached a solution or at the end of the 15 min. In the full study, students received interventions aimed at improving their problem-solving practices and proceeded to solve more black box problems. Here we only consider their performance on the first black box problem before any intervention. Data collected on participants' problem-solving performance includes (1) their solutions (circuit diagram) to the hidden circuit; (2) video recordings of their problem-solving processes; (3) log data recording participants' interactions with the task environment in JavaScript Object Notation (JSON) files as they worked on the problem.

Coding We devised a rubric to score the diagrams submitted by students for the hidden circuit in three dimensions: circuit structure, electrical components, and values of the components (Salehi, 2018). Each dimension has a score between 0 and 2, making the total solution score ranging from 0 to 6. Students' problem-solving outcomes were classified based on their solution scores into three levels: high performing (a score of 5 or 6), medium (3 or 4), and low (0, 1, or 2).

We used a separate rubric to score the effectiveness of students' problem-solving practices based on the video recordings of their problem-solving processes and think-aloud protocols (Salehi, 2018). The rubric was developed by the main researcher who has backgrounds in both electrical engineering and education. The specific practices scored by the rubric include problem definition, decomposition, data collection, data recording, and reflection on solution. These practices were evaluated on a four-point scale, ranging from not effective at all (0) to highly effective (3). Two researchers independently coded 20% of the video recording data to verify the reliability of the rubric and reached agreement for at least 80% of the coded instances for each problem-solving practice. The practice scores provide a baseline measure of individual students' effectiveness at adopting specific problem-solving practices, thus allowing us to evaluate how distinct features extracted from log data correspond to these specific practices.

3.2 *Experiment 2*

Materials The mystery gift problem in the PhET Balancing Act simulation asks students to figure out the mass of a gift using bricks of known weights and a beam that rotates around its center (Fig. 3.2). Students can place bricks and the mystery gift at various marked locations on the beam in the "Setup" mode and observe the outcome of the setup (i.e., how the beam would rotate or stay balanced) in the "Test" mode. The simulation does not allow bricks to be stacked on top of each other, making the problem less intuitive and more difficult. Furthermore, the mass of the mystery gift was deliberately chosen to be unsolvable using a single brick. Balancing the beam thus requires a combination of different bricks placed at various locations on the beam.

Like the black box problem, the mystery gift problem exhibits characteristics of authentic problems in science and engineering domains. Solving the mystery gift requires applying physics knowledge, the torque formula in this case. The problem also provides no data upfront, and students have to collect the data needed for

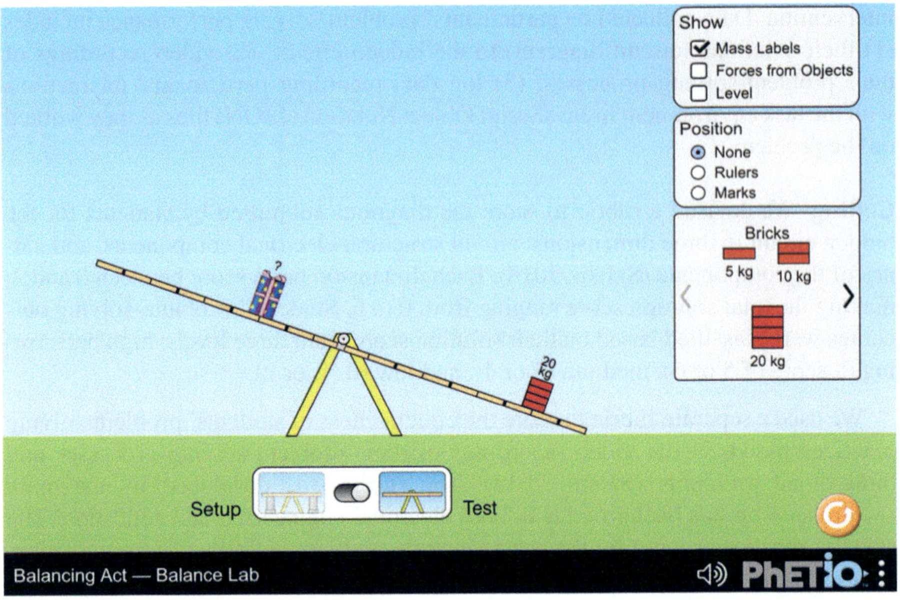

Fig. 3.2 The mystery gift problem asks students to determine the mass of a mystery gift using a balance scale and bricks with known weights. (Image by PhET Interaction Simulations, licensed under CC-BY 4.0)

solving the problem by deciding where to place the mystery gift and bricks on the beam.

Participants Eighty undergraduate students in STEM majors in the United States (48% female) were recruited via an online research crowdsourcing platform, Prolific (Palan & Schitter, 2018). Participants completed the online study at a time and location of their choice. They were compensated for their participation and had the opportunity to get a bonus for correctly solving the problem. Four participants were excluded from the data analysis due to extremely low time-on-task (<1 min). A fifth participant was excluded for missing log data.

Procedures After going through the consent form and a brief tutorial on how the Balancing Act simulation works, participants worked on the first mystery gift problem with the goal of solving it in 15 trials or less. Participants then viewed a worked example demonstrating how to solve the problem and continued to solve a second mystery gift problem afterward. In this chapter, we focus our analysis on participants' performance in the first problem prior to any interventions. Data collected on participants' problem-solving performance includes (1) their solutions for the mystery gift (the gift's mass in kg); (2) their responses to a post-task question probing whether the torque formula was used to solve the problem; (3) log data recording participants' interactions with the task environment in JSON files.

Coding We classified students' problem-solving outcomes into three levels based on the correctness of their answers. Solving the gift's weight requires students to collect useful data by balancing the gift with bricks and make accurate calculations using the torque formula. Missing either element would prevent a student from obtaining the correct solution. Students in the high-performing group correctly solved the gift's weight. Students in the medium-performing group reached solutions that were within a reasonable range from the correct value (+/−2.5 kg), and students in the low-performing group submitted incorrect and far-off answers. We also coded whether a student took notes/recorded data using a table provided in the task environment during the problem-solving process, and whether a student self-reported applying the torque formula to solve the problem.

3.3 Log Data Processing

In processing the log data of students solving the black box and mystery gift problems, we have to make decisions regarding how to parse a continuous stream of actions and what actions to focus on. We wrote a Python script to parse the JSON log files. Our first goal is to filter the log data to reveal a time-stamped sequence of actions taken by students that a human observer would be able to discern through viewing a video recording of students solving the problem. These actions may include connecting a battery when solving the black box and placing a brick on the beam when solving the mystery gift. However, focusing primarily on discrete actions may not be sufficient for differentiating the strategies and practices used by students to solve problems, as demonstrated by Wang et al. (2021a, b). Therefore, we draw from theories of problem-solving practices and our qualitative analyses of students' work processes to group series of actions into semantically meaningful events. Examples of an event include building a circuit when solving the black box problem and setting up a test trial when solving the mystery gift problem. These events represent a higher level of abstraction of the raw log data than discrete actions and may provide direct evidence of students' underlying problem-solving competency.

Additionally, we calculated the periods of inactivity ("pauses") after specific events as a proxy for the behaviors and cognitive processes that are not directly captured by log data. The nature and context of pauses during students' problem-solving processes may be related to several strategies and practices adopted, such as planning, reflection, as well as working with the data collected offline to solve the problem. To explore the cognitive and metacognitive processes underpinning these pauses, we propose a data-driven approach to distinguish different types of pauses based on their duration as well as the context in which they occur. The pause analysis is further triangulated with video recordings of students' problem-solving activities and think-aloud protocols. We will examine whether different types of pauses are associated with students' problem-solving performance measured by their solution quality. Lastly, the pause-based features, along with other features extracted from the log data, are used in regression models to predict students' effectiveness in specific problem-solving practices.

4 Results

4.1 Problem-Solving Outcomes as Measured by Solution Quality

Both the black box and mystery gift were challenging problems for the college students in our study. For the black box problem, 18% of the students in the sample were in the high-performing group based on their solution scores. 35% were in the medium-performing group, and 47% were in the low-performing group. For the mystery gift problem, 28% of the students were in the high-performing group and correctly solved the mass of the gift as 18 kg. 19% were in the medium-performing group and reached a near-correct solution, while 53% were in the low-performing group and submitted answers that were far off. We also found that students took notes and applied domain knowledge at different rates. 45% of students used a table embedded in the task environment to take notes of the data collected, and 41% reported applying the torque formula when solving the mystery gift problem in the post-task survey.

4.2 Problem-Solving Processes as Captured by Features Extracted from Log Data

While the quality of students' solutions can serve as an outcome measure of their problem-solving performance, log data contains detailed information about the processes students went through to solve the problem. Actions taken by students when solving the black box include adding/removing various electrical components (battery, wire, resistor, and light bulbs) to/from the black box and taking measurements using the voltmeter and ammeter. Actions taken by students when solving the mystery gift include adding/removing various bricks and the mystery gift to/from the balance beam and switching between "Setup" and "Test" to set up a test trial and observe its outcome.

Next, we grouped subsets of actions into meaningful events and calculated the duration of pause after these events. Meaningful events during the solution process of the black box problem include building circuits and taking voltage and current measurements using the voltmeter and ammeter. Students on average built 21 circuits, ranging from 0 to 59 circuits. Meanwhile, students on average used the voltmeter 27 times (range: 0–169) and the ammeter 18 times (range: 0–100).

We further categorized the circuits built by students as either simple or complex: simple circuits connect only two wires ("terminals") of the black box, while complex circuits connect more than two wires at a time (Fig. 3.3). This categorization is guided by our knowledge that building simple circuits reflects an effort to decompose the problem into modularized, easy-to-interpret parts, which is a key dimension in the framework of problem-solving practices. Building simple circuits is also an indicator of effective data collection practice, as it allows for the collection of relevant and easy-to-interpret voltage and current readings. As the black box has four wires ("terminals"), six distinct simple circuits are needed to connect each pair of wires. The voltage and current readings from these simple circuits are useful for pinpointing the type and value of the electrical components in a specific segment of the hidden circuit. On the other hand, building complex circuits reflects poor decomposition and data collection practices, as complex circuits give readings that are hard to interpret. Table 3.1 presents the features extracted from the log data of students solving the black box problem grouped by event type.

We adopted a similar analysis for processing the log data of the mystery gift problem and extracted a time-stamped sequence of test trials set up by students. Each test trial consists of the mystery gift and one or more bricks at different locations on the beam. We also calculated the pause time after individual test trials. Students on average set up 18 trials when attempting to solve the mystery gift, ranging from 4 to 106 trials. We further categorized a trial as either simple or complex based on the total number of objects used in the trial: a simple trial uses no more than three objects on the beam (including the mystery gift and bricks), while a complex trial uses four or more objects (Fig. 3.4). Setting up simple trials is an indicator

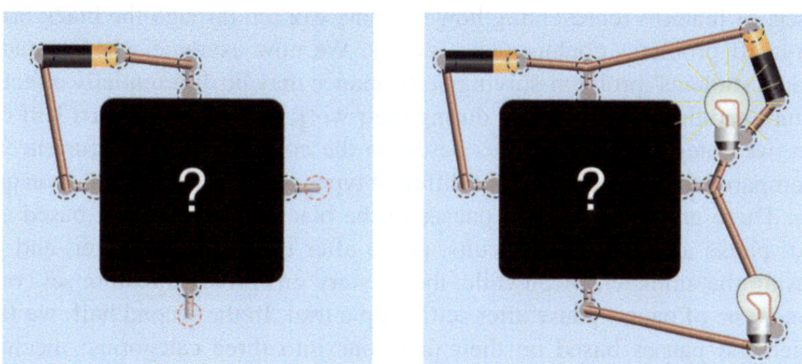

Fig. 3.3 Examples of simple (left) and complex (right) circuits built by students

Table 3.1 Features extracted from the log data of students solving the black box problem

Circuit event	Measurement event	Pause event
Building simple circuits	Using the voltmeter	Pausing after building a circuit
Building complex circuits	Using the ammeter	Pausing after using the voltmeter
		Pausing after using the ammeter

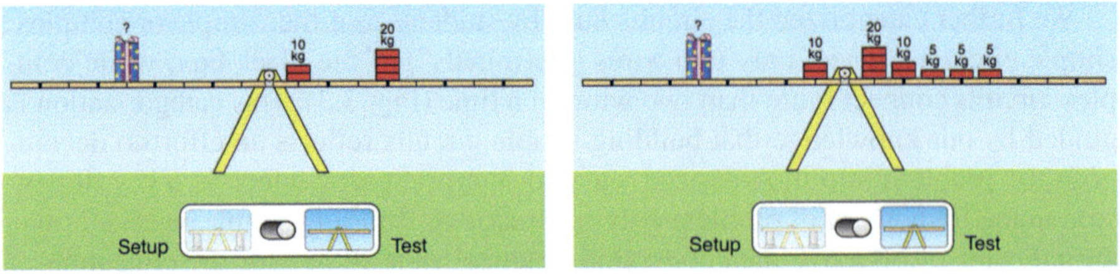

Fig. 3.4 Examples of simple (left) and complex (right) trials set up by students

Table 3.2 Features extracted from the log data of students solving the mystery gift problem

Test trial event	Pause event
Setting up simple test trials	Pausing after a test trial
Setting up complex test trials	

of effective data collection practice, as it yields data that enables fast and easy calculation. In contrast, complex trials would lead to complicated calculations and make it challenging to estimate the weight of the mystery gift. Table 3.2 presents the final set of features extracted from the log data of students solving the mystery gift problem.

4.3 Pause as a Generalizable Indicator of Deliberate Problem Solving

Both sets of features representing how students worked through the black box and mystery gift problems contain pause events. We now examine whether and how individual students' problem-solving performance may be differentially affected by the nature and duration of pauses during their work process. In the first half of this section, we categorized the pauses based on the context of their occurrences. We then compared the mean duration of different types of pauses across solution quality groups. There are three types of pauses in the black box feature set based on the context: pause after building circuits, pause after using the voltmeter, and pause after using the ammeter. Meanwhile, the mystery gift problem feature set contains only one type of pause, pause after setting up a trial. In the second half, we further classified the pauses based on their durations into three categories: mechanical pause, deliberate pause, and distracted pause.

We found that high-performing students on average paused significantly longer than low-performing students after using the ammeter when solving the black box problem. Figure 3.5 shows the boxplots of the mean pause duration of individual students after building circuits (left), using the voltmeter (middle), and using the ammeter (right). We applied one-way ANOVA tests to compare the pause duration across groups with varying levels of solution qualities (high, medium, and low). While there was only a marginal difference in the pause duration post circuit

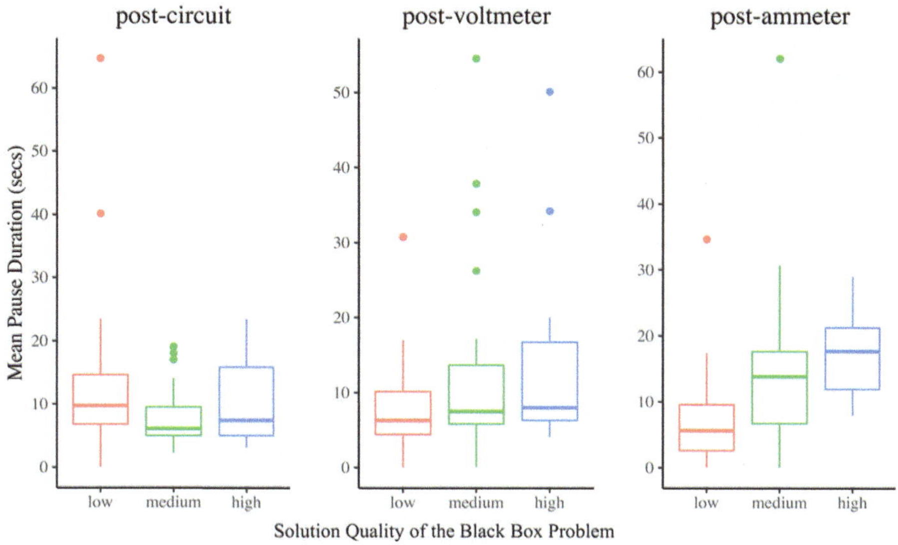

Fig. 3.5 Mean pause durations of individual students grouped by solution qualities of the black box problem

construction [F(2, 69) = 2.26, p = 0.11] and voltmeter usage [F(2, 69) = 2.49, p = 0.09] across the three groups, the pause duration post ammeter usage differed significantly depending on the solution group [F(2, 69) = 8.61, $p < 0.001$]. Post-hoc nonparametric Wilcoxon tests showed that students in the high-performing group on average paused significantly longer than students in the low-performing group after using the ammeter ($p < 0.0001$), and that students in medium-performing group also paused significantly longer than students in the low-performing group ($p < 0.001$). The high-performing group paused longer than the medium-performing group post-ammeter as well, but the difference was not statistically significant ($p = 0.25$).

We found a similar trend of high-performing problem solvers pausing longer when solving the mystery gift problem. Figure 3.6 shows a boxplot of the mean pause duration of individual students after setting up a trial grouped by solution quality. One-way ANOVA test revealed a significant group effect on the duration of pauses [F(2, 72) = 8.03, $p < 0.001$]. Post-hoc Wilcoxon tests showed that students in the high-performing group on average paused significantly longer than those in the low-performing group ($p = 0.01$). The difference between the medium- and low-performing group was marginally significant ($p = 0.06$), while there was no significant difference between the high- and medium-performing groups.

Next, we categorize pauses into three categories after inspecting the distributions of all pause durations for evidence of mixture distributions and reviewing the video recordings of students' problem-solving processes. The first type of pause follows a relatively normal distribution that ranges from 0 to 9 s with a mean of around 3 or 4 s. We label these short pauses as *mechanical pauses*. These pauses constitute the

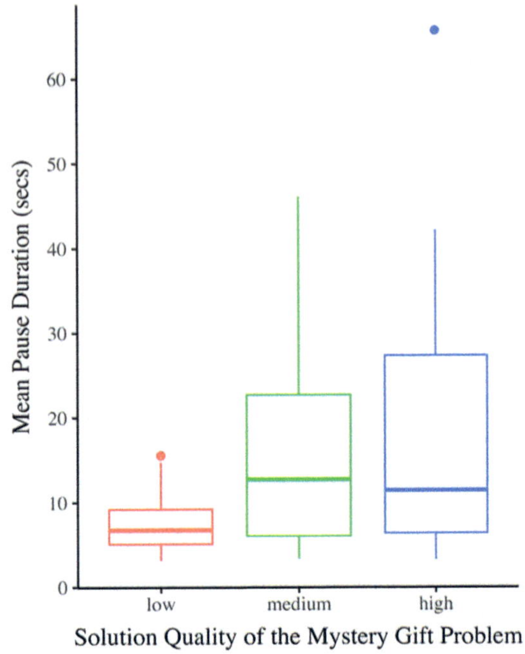

Fig. 3.6 Mean pause durations of individual students grouped by solution qualities of the mystery gift problem

time it takes to view an animation or move the cursor around in the simulations with minimal cognitive processing of the information just obtained. The second type of pause is longer than 10 s and represents a conscious and deliberate effort of stepping back from interacting with the simulation to make progress in solving the problem. We label these longer pauses as *deliberate pauses*. Based on our qualitative observations of students' problem-solving processes, we found that problem-solving practices adopted during a deliberate pause might include taking notes of the data collected (data recording), applying domain knowledge and doing calculations to make sense of the data collected (data interpretation), and summarizing the progress made so far and revising ineffective strategies if necessary (reflection). The third type of pause, distracted pauses, is an outlier in terms of length and is longer than 3 min. Possible origins of distracted pauses include technical glitches of the task environment, communications with the researcher in lab settings, and off-task behaviors in online settings where students worked on their own without any supervision. These cut-offs (10 s & 3 min) are arbitrary and do not take into account the variations in individual students' cognitive processing speeds or the difficulty levels of the tasks. Nonetheless, they provide a useful approach for identifying the meaningful pauses from the log data of students' interactions with OELE-based tasks that warrant further investigation.

The percentage of deliberate pauses of individual students was positively associated with their problem-solving success. Figure 3.7 shows the mean relative frequency of different types of pauses across the three solution quality groups. Distracted pauses were excluded from the plots due to their infrequent occurrence:

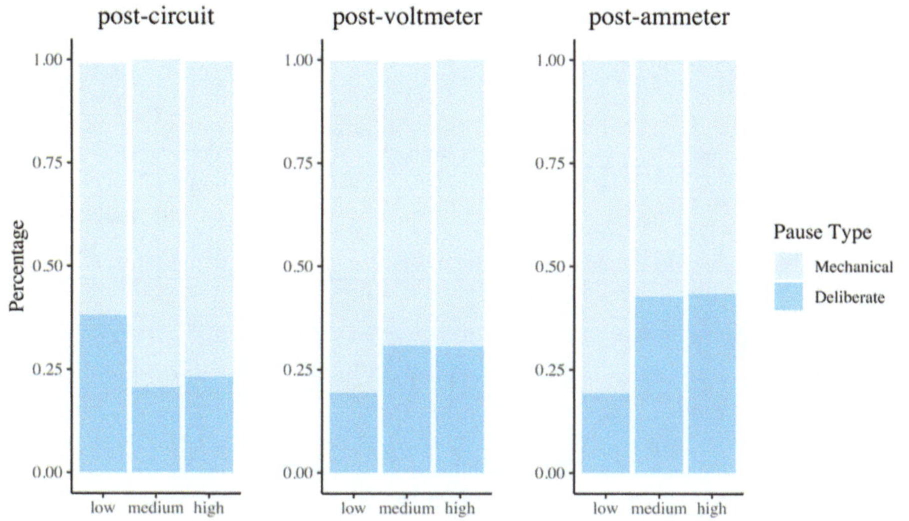

Fig. 3.7 Mean relative frequency of mechanical vs. deliberate pauses of individual students when solving the black box problem

there were a total of six instances of distracted pause belonging to six students as captured by the black box log data (0.14% of all pauses), and two instances of distracted pause belonging to two students as captured by the mystery gift log data (0.15% of all pauses). One-way ANOVA tests indicated that there was a significant group effect on the percentage of deliberate pauses post-circuit [$F(2, 69) = 3.44$, $p < 0.05$] and post-ammeter [$F(2, 69) = 8.03$, $p < 0.001$]. A post-hoc Wilcoxon test revealed that students in the low-performing group had a significantly higher percentage of deliberate pauses after building circuits than students in the medium-performing group ($p < 0.01$). This direction was reversed in the pauses after using the ammeter. Low-performing students had a significantly lower percentage of deliberate pauses than both the medium-performing students (Wilcoxon test, $p < 0.001$) and the high-performing ones ($p < 0.0001$).

Similarly, students who successfully solved the mystery gift problem had a higher portion of deliberate pauses. Figure 3.8 presents the mean relative frequency of different types of pauses derived from the log data of the mystery gift problem. We found a significant group effect [$F(2, 72) = 9.70$, $p < 0.001$] on the percentage of deliberate pauses. Students in the low-performing group had a significantly lower percentage of deliberate pause than those in the medium-performing (Wilcoxon test, $p < 0.05$) and high-performing group ($p < 0.01$).

To summarize, our analyses suggest that successful problem solvers paused longer on average and had a higher percentage of deliberate pauses. This trend was observed in students' solution processes of both the black box and mystery gift problems. These longer, deliberate pauses represent students' efforts to step back from interacting with the task environment in order to make progress in solving the

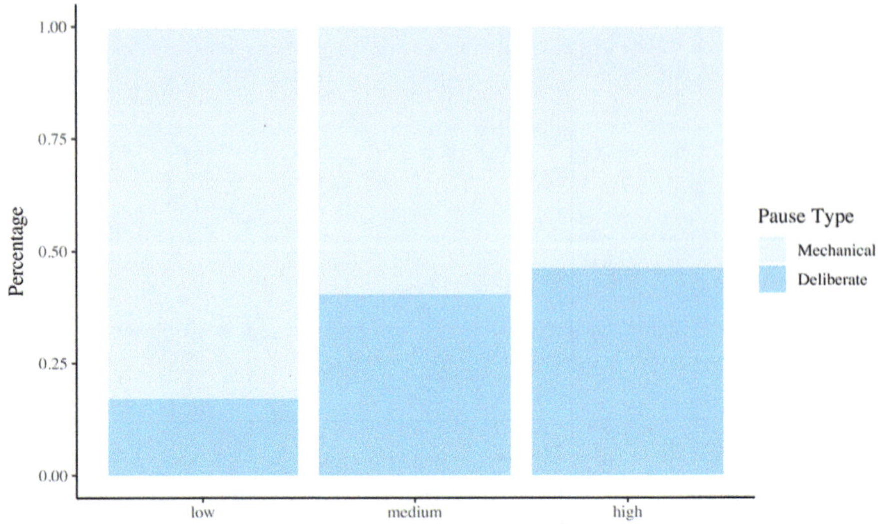

Fig. 3.8 Mean relative frequency of mechanical vs. deliberate pauses of individual students when solving the mystery gift problem

problem. Potential problem-solving practices adopted during these deliberate pauses include data recording, data interpretation, as well as reflection.

The difference in pauses between high- and low-performers was also reflected in the contexts when these pauses occurred. In the case of the black box problem, two sets of data, voltage and current, need to be collected from each segment of the hidden circuit in order to infer the electrical components in the specific segment. The most expert-like solution path would be to first measure the voltage to check if the hidden segment contains a battery, followed by adding an external battery in the simple circuit when necessary and measuring the current using the ammeter. This allows for calculating the resistance in the circuit using Ohm's law, or $R = V/I$. Given the order of effective data collection (voltage first and current second), it is not surprising that the difference in pause duration between high- and low-performing students was most pronounced after the ammeter usage. On the other hand, longer pauses after building circuits may indicate that a problem solver attempted to make sense of a signal, such as a bulb lighting up, yet were largely unsuccessful due to the lack of precision in the signal.

4.4 How Log Data-Based Features Were Associated with Specific Problem-Solving Practices

We now turn to examine how pauses and other log data-based features were associated with the effectiveness of specific problem-solving practices as hand scored by researchers. Multivariate linear regression models were applied to map the features extracted from log data onto students' problem-solving practice scores. We built and evaluated the models using the caret package in the R statistical programming environment (Kuhn, 2008). Model performance was evaluated by the R-squared metrics obtained through five-fold cross-validation.

Results of the regression analyses showed that features extracted from the log data accounted for a large fraction of the variance in the researcher-coded scores of specific problem-solving practices. Table 3.3 presents a summary of the regression models predicting students' effectiveness in decomposition, data collection, and data recording when solving the black box problem. Students' scores of the decomposition practice were closely associated with the percentage of complex circuits: for each unit increase in the percentage of complex circuits built, the decomposition score would go down by close to 0.70 units. Meanwhile, scores measuring students' data collection effectiveness were significantly predicted by the number of distinct simple circuits and the percentage of complex circuits built. Simple circuits built to connect an additional pair of terminals of the black box were associated with a 0.34 unit increase in the data collection score, while each unit increase in the percentage of complex circuits was associated with a 0.29 decrease in data collection effectiveness. Lastly, the scores measuring students' effectiveness in data recording were closely associated with both the number of distinct simple circuits and the percentage of deliberate pause after ammeter usage. Simple circuits built to connect an additional pair of terminals would increase the data recording score by 0.20 units, and a one-unit increase in the percentage of deliberate pauses would achieve a similar effect on the data recording score.

For the mystery gift problem, we didn't manually score students' effectiveness in specific problem-solving practices as their work processes were not audio- or

Table 3.3 Linear regression models predicting problem-solving practice scores using log data-based features of students solving the black box problem

	Problem-solving practices as outcome variables coefficient (SE)		
Features extracted from log data	Decomposition	Data collection	Data recording
Total circuit count	−0.17 (0.09) .	−0.11 (0.10)	−0.03 (0.12)
Distinct simple circuit	0.13 (0.07) .	**0.34 (0.07)*** **	**0.20 (0.09)***
% of complex circuits	**−0.69 (0.10)*****	**−0.29 (0.11)***	−0.12 (0.13)
Voltmeter count	0.07 (0.07)	−0.02 (0.08)	−0.09 (0.09)
Ammeter count	0.03 (0.08)	0.04 (0.09)	0.19 (0.10) .
% of deliberate pauses post circuits	−0.10 (0.08)	0.03 (0.09)	−0.10 (0.11)
% of deliberate pauses post voltmeter	0.04 (0.07)	0.06 (0.08)	0.19 (0.10) .
% of deliberate pauses post ammeter	−0.09 (0.08)	−0.06 (0.09)	**0.21 (0.11)***
R-squared (Five-fold cross validation)	**0.70**	**0.63**	**0.50**

***$p < 0.001$, **$p < 0.01$, *$p < 0.05$, . $p < 0.1$

video-recorded in the online study. Instead, we coded students' data recording practice as 0/1 by checking whether they recorded any data in the table provided in the task environment and coded the data interpretation practice in a similar fashion by checking whether students self-reported applying physics knowledge (torque) when solving the problem. We found that individual problem solvers' percentage of deliberate pauses was significantly correlated with the adoption of data recording (r_{pb} = 0.55, $p < 0.001$), as well as with effective data interpretation (r_{pb} = 0.34, $p < 0.01$).

Results of this study provide support for applying data mining techniques to log data generated in OELEs to automate the assessment of students' higher-order competencies such as problem-solving practices. When a new group of students works on the black box problem, we would be able to estimate their effectiveness in decomposition, data collection, and data recording with reasonable accuracy using log data alone in an automated and efficient workflow. At the same time, the regression models using log data-based features did not account for substantial variances in the other two problem-solving practices coded by researchers: problem definition and reflection on solution. These two practices were evaluated primarily based on students' think-aloud utterances at the beginning and the end of the problem-solving process, respectively. Automating the assessment of problem definition and reflection on solution practices may require us to explore additional machine learning techniques such as natural language processing (NLP) and modifications to the task environment to allow for problem solvers to verify their proposed solutions.

5 Discussion

This study demonstrates the usefulness of log data in providing unobtrusive observations of students' work processes in solving interactive, authentic problems and the value of human knowledge in guiding feature extraction from log data. Results show that semantically meaningful events initiated by students during problem solving, including constructing circuits, taking measurements, setting up test trials, and pausing, can serve as significant predictors of their effectiveness in specific problem-solving practices as well as general problem-solving success. In particular, pauses were found to be significantly correlated with students' problem-solving performance. The longer the students paused after specific meaningful events, the more likely they were to solve the problems. Additionally, these pauses were significant predictors of students' effectiveness in data recording and data interpretation.

The present study extends previous research work using log data to capture and evaluate students' competence in problem solving. Features extracted from log data have been linked to students' exploration strategies such as vary-one-thing-at-a--time (VOTAT) and designing controlled experiments (Gobert et al., 2013; Greiff et al., 2015; Käser & Schwartz, 2020). Results from the present study revealed that problem-solving practices such as decomposition, data collection, data recording, and data interpretation were also conducive to being measured by the log data generated in interactive, knowledge-rich problems. For the black box problem, the percentage of complex circuits built was negatively correlated with students' effectiveness in decomposition and data collection, while the number of distinct simple circuits built was positively correlated with students' effectiveness in data collection and data recording. These circuit-type features were derived from the raw log data through a theory-driven approach that incorporates domain experts' knowledge into the feature extraction process.

Our analyses identified pause as a shared key feature derived from the log data of two distinct problem-solving tasks completed by two groups of students in different settings (in-person lab and online study). Results showed that the nature of a pause during students' problem-solving process was determined by both its duration and the context of its occurrence. We classified pauses into three types based on their durations: (1) short, mechanical pauses encompassing the time it took to view an animation or move around in the task environment, (2) deliberate pauses representing the efforts of stepping back from interacting with the task to work with the data collected and evaluate problem-solving progress, and (3) distracted pauses indicating the occurrence of off-task behaviors. We found that problem solvers who obtained the correct solutions on average paused longer and had a higher percentage of deliberate pauses than those who only reached solutions that were far off. For the black box problem, the differences in pause duration and composition were most pronounced for the pauses after ammeter usage but not significant for the pauses after voltmeter usage, reflecting differentiated data collection paths between high- and low-performing problem solvers. A higher percentage of deliberate pauses after ammeter usage was an indicator of students' effectiveness in data recording when solving the black box problem. Similarly, the percentage of deliberate pauses was positively correlated with students' effectiveness in data recording and data interpretation practices when solving the mystery gift problem.

Results from the present study revealed the complex nature of pauses taken by students while working on OELE-based tasks and extended previous studies studying the phenomenon. For instance, Gobert et al. (2015) identified long pauses from interacting with the Inq-ITS simulation as signaling disengagement from the task goal. In contrast, Perez et al. (2017) found that high-performing students took more pauses that were longer than 15 s after testing circuits than low-performing students in an inquiry task embedded in PhET simulation. Similarly, Bumbacher et al. (2018) highlighted sufficiently long pauses between experiments as an indicator of deliberate planning and reflection in inquiry-based learning tasks. These divergent findings can be reconciled by considering pause during students' work processes as a multifaceted rather than monolithic construct. Future research using log data of students' work processes to model high-level competencies should extract pause as an important feature and investigate the underlying cognitive and metacognitive processes associated with different types of pauses depending on their contexts and durations.

This study also carries important implications for future development of assessment tasks embedded in interactive learning environments. First, the study shows that it is possible to capture and measure students' competencies in problem-solving practices using the logged interaction data of students' work processes, especially their practices related to decomposition and data collection. Furthermore, the results demonstrate the value of incorporating expert knowledge into feature extraction (feature engineering) to identify semantically meaningful behavioral patterns from the raw log data. Feature engineering should be an important consideration to improve the predictive power of models in the field of educational data mining. The main reason is that datasets of students working with OELE-based tasks are generally not large enough to enable the training of deep machine learning models that can automatically discover relevant features from the raw data. Lastly, these features offer the basis for providing adaptive and timely feedback for students to iterate and improve their problem-solving practices as they work on the task.

6 Limitations

Both problem-solving tasks employed in the present study require knowledge of physics (mechanics and electricity). Future research is needed to establish the generalizability of the problem-solving practices identified in this study and understand how these practices manifest in interactive tasks from other STEM domains. Furthermore, the features presented in this study represent only a subset of features that could be extracted from the log data. It is possible that the most predictive feature is yet to be discovered. Nonetheless, our results present a promising direction for leveraging log data to automate the assessment of high-level competencies.

7 Conclusion

The study presents empirical evidence that log data generated in OELE-based tasks can be used to measure high-level constructs like problem-solving practices through the extraction of both task-specific and task-general features. Specifically, pausing deliberately after data collection was an important indicator of general problem-solving success as well as effectiveness in specific problem-solving practices. This is an important result, as it reveals a potentially generalizable pattern of how students interact with digital learning environments. The dynamic and instantaneous interactivity of the digital platform can make it enticing for students to act in a fast-paced, trial-and-error manner without thinking deeply about the information gathered or the learning goals. Teaching students to adopt deliberate pauses when working with OELE-based tasks could lead to more effective problem-solving practices and should be an important goal for researchers, teachers, and designers of digital learning technologies.

Acknowledgments This research was supported by the Gordon and Betty Moore Foundation.

References

ABET Engineering Accreditation Commission. (2022). *Criteria for accrediting engineering programs*. ABET.

Bumbacher, E., Salehi, S., Wieman, C., & Blikstein, P. (2018). Tools for science inquiry learning: Tool affordances, experimentation strategies, and conceptual understanding. *Journal of Science Education and Technology, 27*(3), 215–235.

Csapó, B., & Funke, J. (2017). The development and assessment of problem solving in 21st-century schools.

De Jong, T., & Van Joolingen, W. R. (1998). Scientific discovery learning with computer simulations of conceptual domains. *Review of Educational Research, 68*(2), 179–201.

Fischer, C., Pardos, Z. A., Baker, R. S., Williams, J. J., Smyth, P., Yu, R., et al. (2020). Mining big data in education: Affordances and challenges. *Review of Research in Education, 44*(1), 130–160.

Fratamico, L., Conati, C., Kardan, S., & Roll, I. (2017). Applying a framework for student modeling in exploratory learning environments: Comparing data representation granularity to handle environment complexity. *International Journal of Artificial Intelligence in Education, 27*(2), 320–352.

Gašević, D., Greiff, S., & Shaffer, D. W. (2022). Towards strengthening links between learning analytics and assessment: Challenges and potentials of a promising new bond. *Computers in Human Behavior*, 107304.

Gobert, J. D., Sao Pedro, M., Raziuddin, J., & Baker, R. S. (2013). From log files to assessment metrics: Measuring students' science inquiry skills using educational data mining. *Journal of the Learning Sciences, 22*(4), 521–563.

Gobert, J. D., Baker, R. S., & Wixon, M. B. (2015). Operationalizing and detecting disengagement within online science microworlds. *Educational Psychologist, 50*(1), 43–57.

Greiff, S., Wüstenberg, S., & Avvisati, F. (2015). Computer-generated log-file analyses as a window into students' minds? A showcase study based on the PISA 2012 assessment of problem solving. *Computers & Education, 91*, 92–105.

Hannafin, M. J., & Land, S. M. (1997). The foundations and assumptions of technology-enhanced student-centered learning environments. *Instructional Science, 25*(3), 167–202.

Holthuis, N., Deutscher, R., Schultz, S. E., & Jamshidi, A. (2018). The new NGSS classroom: A curriculum framework for project-based science learning. *American Educator, 42*(2), 23–27.

Kardan, S., & Conati, C. (2011). A framework for capturing distinguishing user interaction behaviors in novel interfaces. In *EDM* (pp. 159–168).

Käser, T., & Schwartz, D. L. (2020). Modeling and analyzing inquiry strategies in open-ended learning environments. *International Journal of Artificial Intelligence in Education, 30*(3), 504–535.

Kuhn, M. (2008). Building predictive models in R using the caret package. *Journal of Statistical Software, 28*, 1–26.

Land, S., & Jonassen, D. (2012). *Theoretical foundations of learning environments*. Routledge.

Levy, F., & Murnane, R. J. (2013). *Dancing with robots: Human skills for computerized work*. Third Way NEXT.

National Research Council. (2011). *Learning science through computer games and simulations*. National Academies Press.

Newell, A., & Simon, H. A. (1972). *Human problem solving* (Vol. 104, No. 9). Prentice-Hall.

NGSS Lead States (2013). Next Generation Science Standards: For States, By States. Washington, DC: The National Academies Press.

Organisation for Economic Co-operation and Development (OECD). (2005). *Problem solving for tomorrow's world: First measures of cross-curricular competencies from PISA 2003*. PISA, OECD Publishing.

Organisation for Economic Co-operation and Development (OECD). (2014). *PISA 2012 results: Creative problem solving: Students' skills in tackling real-life problems* (Vol. V). OECD Publishing.

Palan, S., & Schitter, C. (2018). Prolific. ac – A subject pool for online experiments. *Journal of Behavioral and Experimental Finance, 17*, 22–27.

Pedaste, M., Mäeots, M., Siiman, L. A., De Jong, T., Van Riesen, S. A., Kamp, E. T., et al. (2015). Phases of inquiry-based learning: Definitions and the inquiry cycle. *Educational Research Review, 14*, 47–61.

Perez, S., Massey-Allard, J., Butler, D., Ives, J., Bonn, D., Yee, N., & Roll, I. (2017, June). Identifying productive inquiry in virtual labs using sequence mining. In *International conference on artificial intelligence in education* (pp. 287–298). Springer.

Polya, G. (1971). *How to solve it: A new aspect of mathematical method* (Vol. 85). Princeton University Press.

Price, A., Salehi, S., Burkholder, E., Kim, C., Isava, V., Flynn, M., & Wieman, C. (2022). An accurate and practical method for assessing science and engineering problem-solving expertise. *International Journal of Science Education*, 1–24.

Salehi, S. (2018). *Improving problem-solving through reflection*. Stanford University.

Stadler, M., Herborn, K., Mustafić, M., & Greiff, S. (2020). The assessment of collaborative problem solving in PISA 2015: An investigation of the validity of the PISA 2015 CPS tasks. *Computers & Education, 157*, 103964.

Wang, K. D., Salehi, S., Arseneault, M., Nair, K., & Wieman, C. (2021a, June). Automating the assessment of problem-solving practices using log data and data mining techniques. In *Proceedings of the eighth ACM conference on learning@ scale* (pp. 69–76).

Wang, K., Nair, K., & Wieman, C. (2021b, April). Examining the links between log data and reflective problem-solving practices in an interactive task. In *LAK21: 11th international learning analytics and knowledge conference* (pp. 525–532).

Wang, K. D., Cock, J. M., Käser, T., & Bumbacher, E. (2023). A systematic review of empirical studies using log data from open-ended learning environments to measure science and engineering practices. *British Journal of Educational Technology, 54*(1), 192–221.

Wieman, C. E., Adams, W. K., & Perkins, K. K. (2008). PhET: Simulations that enhance learning. *Science, 322*(5902), 682–683.

Windschitl, M., Thompson, J., & Braaten, M. (2008). Beyond the scientific method: Model-based inquiry as a new paradigm of preference for school science investigations. *Science Education, 92*(5), 941–967.

Wu, M., & Adams, R. (2006). Modelling mathematics problem solving item responses using a multidimensional IRT model. *Mathematics Education Research Journal, 18*(2), 93–113.

Chapter 4
Challenges in Assessments of Soft Skills: Towards Unobtrusive Approaches to Measuring Student Success

Abhinava Barthakur, Vitomir Kovanovic, Srecko Joksimovic, and Abelardo Pardo

Abstract Rapid technological advances, coupled with globalization, have resulted in a changing economy, requiring graduates and students to master not only technical and subject knowledge but also broad, transferable skills for workplace readiness. However, assessing these essential soft skills and competencies beyond the cognitive domain has often relied on questionnaires, surveys and other self-rated scales, which are subjective, often obtrusive in nature, subject to response biases, and lack scalability. In contrast, the pervasive use of educational technology has provided researchers with the opportunity to unobtrusively collect enormous amounts of factual learners' data which has the potential to overcome some of the challenges with questionnaire-based approaches. These unobtrusive measures increase the possibilities of passively evaluating skill acquisition and supporting learners by personalizing learning according to their needs. This chapter outlines a multi-tiered case study and proposes a novel blended methodology, marrying measurement models and learning analytics techniques to mitigate some of these challenges and unobtrusively measure leadership skills in a workplace learning context. Using learners' reflection assessments, several leadership-defining course objectives were quantified, and their progress was assessed over time. The implications of this evidence-based assessment approach, informed by theory, to measure and model soft skills acquisition are further discussed.

Keywords Assessment · Soft skills · Learning analytics · Measurement theory · Online learning

A. Barthakur (✉) · S. Joksimovic · A. Pardo
University of South Australia, Adelaide, SA, Australia
e-mail: abhinava.barthakur@unisa.edu.au; srecko.joksimovic@unisa.edu.au; abelardo.pardo@unisa.edu.au

V. Kovanovic
Centre for Change and Complexity in Learning, University of South Australia, Adelaide, SA, Australia
e-mail: vitomir.kovanovic@unisa.edu.au

V. Kovanovic et al. (eds.), *Unobtrusive Observations of Learning in Digital Environments*, Advances in Analytics for Learning and Teaching,
https://doi.org/10.1007/978-3-031-30992-2_4

1 Introduction

The changing nature of the modern workplace coupled with technological advances has led to short shelf life for educational practices focused on rote learning within traditional settings, shifting the focus on learners applying their skills and knowledge (Pellegrino, 2017) and on assessment of competencies (Milligan, 2020). Unlike classroom activities that are often constrained to independent learning within disciplinary boundaries, real-world challenges in the modern workplace typically demand collaborative and individual learning approaches to transfer theoretical understandings to practical applications (Ginda et al., 2019). Business leaders, employers, educational stakeholders, and researchers recognize that school success is not the only influential factor determining the economy's success (Kyllonen, 2012) and have called for policies that would support and promote the development of more broad, transferable skills. Organizations deem these broad skills essential for workplace success to solve real-world challenges. Researchers and practitioners lack consensus on the terminology describing these skills (Joksimovic et al., 2020), but they are commonly referred to as 21st century competencies or soft skills (Vockley, 2007). Often differing from context to context, soft skills are well accepted as inherently social and developed through collaboration and networking (Jenkins et al., 2006). Specifically, skills such as communication (oral and written), critical thinking, leadership, problem-solving, and teamwork are some of the most desirable competencies for future graduates (Casner-Lotto & Barrington, 2006; Lai & Viering, 2012).[1]

Despite much work being undertaken on promoting soft-transferable skills, they are inherently complex and assessing them is less straightforward (Joksimovic et al., 2020). Martin et al. (2016) note that "even with increasing attention to the importance of 21st century skills, there is still relatively little known about how to measure these sorts of competencies effectively" (*ibid.*, p. 37). The evaluation of the development and acquisition of soft skills is primarily done using introspective approaches such as self-reported questionnaires and inventories (Amagoh, 2009; Ebrahimi & Azmi, 2015), often considered obtrusive. Lai and Viering (2012), and Sondergeld and Johnson (2019), among others, note that the cost-effectiveness, ease of implementation and the ability to provide scores on multiple abilities simultaneously make such approaches popular. However, there are several important challenges associated with these commonly adopted introspective measures, including response biases (Bergner, 2017; Gray & Bergner, 2022), scalability and coverage (Pongpaichet et al., 2022), to name a few. Such challenges make the adoption of such introspective approaches much more challenging, prompting the need to look for alternative approaches.

In contrast, adopting online learning platforms along with educational technologies such as a learning management system provides unobtrusive ways of collecting

[1] For convenience and to minimize the multiplicity of terms used to describe the same skills and competencies, we refer to them as soft skills throughout this chapter.

educational data. Relatively recently, educational organizations have started investing in longitudinal learning data, collected at various levels of granularity, to provide insights into teaching quality and understanding the learning process (Joksimovic et al., 2019). The emergence of educational research domains such as *learning analytics* (LA) has demonstrated the potential to support the assessment of learners' skill acquisition through unobtrusive approaches by utilizing fine-grained data collected from educational technologies. Collective sets of LA research in the domain of soft skills measurement have been published in the Journal of Learning Analytics across two editions in 2016 (Shum & Crick, 2016) and 2020 (Joksimovic et al., 2020).

In this chapter, we discuss the use of unobtrusive measures to assess soft skills and illustrate a case study on assessing leadership skills. While other skills, such as creativity, critical thinking, and complex problem-solving, received significant attention, there has been little work on assessing leadership skills. In the next section of the chapter, we discuss the various modalities aimed at developing soft skills, some of the challenges associated with the current assessment approaches and, subsequently, the need for more advanced methodologies. To address these challenges, we outline a case study that measured leadership skills across a MOOC study program in three components. In the first component, we rely on an automated machine learning classifier to extract unobtrusive measures from reflective artefacts. We then explore the use of learning analytics techniques and measurement models to evaluate the mastery and acquisition of leadership in a single MOOC. In the final component of the case study, we explore the systematic longitudinal progression of learners developing their skills. We aim to identify some relationships between the learning objectives across multiple courses and how fulfilling the prerequisites in one course helps learners progress in subsequent courses.

2 Background

2.1 Developing Soft Skills

While most jobs nowadays demand a broad set of skills to adequately deal with real-world challenges and prepare for an unknown future (Rios et al., 2020), the development of such skills and competencies has been an increasing concern among employers and educators (Shum & Crick, 2016; Haste, 2001). For instance, the Partnership for 21st Century Skills (P21) highlighted higher education's inefficiency in adequately developing these transferable skills (Casner-Lotto & Barrington, 2006), resulting in significant issues with graduate employability. Employers and private organizations encouraged universities and educational institutions to incorporate such skills into their curricula and put greater emphasis on developing complex skills. Additionally, the curricula must be constantly re-evaluated and revised depending on the labour market requirement.

Besides developing soft skills within the traditional classroom settings, workplace learning programs are used to develop soft skills to meet the rapid changes in the modern workforce. Organizations worldwide, for example, are developing professional training courses to deliver the skills and competencies required to tackle the ever-changing work demands (Amagoh, 2009; Burke & Collins, 2005). The need for rapidly changing skill sets and new technological affordances have provided the scope for shifting the focus from classroom learning toward leveraging online settings for developing the necessary workforce skills and professional development within their employees. While some organizations encourage employees to acquire job-relevant skills through off-the-shelf courses, others co-create certification courses along with educational providers to reduce the gap between the skills graduates and employees need to be successful in the modern workforce (Ginda et al., 2019). Moreover, unlike traditional courses in higher education, workplace training usually prioritizes learning processes that focus on transferring content knowledge to practical workplace applications.

Online learning has been increasingly seen as a prominent approach to delivering these workforce programs dedicated to upskilling their employees. These trends have also been accelerated by the recent COVID-19 pandemic, which put online learning at the centre of the educational policy of many governments around the world. One modality of online learning that witnessed growing interest in the domain of professional development is Massive Open Online Courses (MOOCs). Besides providing opportunities for gaining conceptual hold over subject-related knowledge, MOOCs, through their varying pedagogy and self-regulated learning, provide learners with opportunities for developing soft-transferable skills for lifelong learning (Chauhan, 2014). The underlying impact of MOOCs in nurturing these highly valued skills in the labour market allows learners to cultivate knowledge and skills beyond a specific domain. Therefore, their use holds great value from not only developing these skills among learners, but also providing unobtrusive means to collect data.

2.2 Leadership Skills

Numerous skills fall under the umbrella of *soft skills*. Although all these skills are indicative of being effective in dealing with challenges within professional life, Rios et al. (2020) argue that employers do not deem all of them equally essential for their organization. Skills such as written communication, deemed critical for any workplace setting, are missing in 47% of 2-year and 28% of 4-year graduates (Casner-Lotto & Barrington, 2006). In contrast, some 21st century competencies, such as social responsibility, are rarely mentioned in job advertisements (Rios et al., 2020). Therefore, the distinction between the development of novel skills and those adjudged necessary ought to lead educators and policymakers to make educational reforms to decrease learning disparities and improve workforce readiness.

One such essential soft skill that is widely accepted in creating organisational impact and increasingly seen as employment quality is leadership capability (Rohs & Langone, 1997). Leadership skills are considered essential by almost 82% of organizations (Casner-Lotto & Barrington, 2006). Leadership skills contribute to a positive work environment and job satisfaction among employees (Amagoh, 2009). As such, to support the development of leadership skills, various instructional programs are offered in both academic and informal workplace learning settings. Along with providing opportunities to enhance problem-solving capabilities, communication and collaboration, these programs facilitate learning through open-ended and unstructured learning tasks (Joksimovic et al., 2020). Another key aspect of these workplace programs is the emphasis on reflection-promoting activities, encouraging participants to reflect on their learnings and professional experiences (Amagoh, 2009; Burke & Collins, 2005). Such reflective practices show potential in continuously developing skills through purposeful consideration of key concepts and transferring knowledge to real work-life scenarios (Helyer, 2015). Therefore, reflection activities are common educational practices that are used as means to measure the growth and acquisition of skills.

2.3 *Challenges of Assessing Soft Skills*

The widely adopted P21 framework of 21st century skills have emphasized the need for assessing the learning and acquisition of soft skills to provide formative intervention to steer and support students' performance (Casner-Lotto & Barrington, 2006). Although the various frameworks developed to understand soft skills provide preliminary empirical evidence of their meaning and value (Pellegrino, 2017), unlike measuring "content" or discipline-specific knowledge in classroom settings, assessing soft skills is far more complex and has been of increasing concern for a couple of reasons: there is a lack of coherent understanding of the nature and development of soft skills (Care et al., 2018) and thus, it is hard to quantify them (Joksimovic et al., 2020). Henceforth, researchers have raised several concerns regarding their measurement. Some of the major concerns associated with the assessments of soft skills are as follows:

- *Biases* – Recruiting learners to participate in self-reported scales includes different response biases (Bergner, 2017; Gray & Bergner, 2022). Some of the commonly observed biases are response shift bias (shift in the frame of reference of the measured construct; Barthakur et al., 2022a, b, c; Rohs & Langone, 1997), social desirability bias (rejecting undesirable characteristics and faking socially desirable traits; Nederhof, 1985), biases that result from participants resorting to extreme ends of Likert scale (Bachman & O'Malley, 1984), among others. As such, although it is assumed that participants are honest while answering these surveys and questionnaires, they are replete with biases that cannot be ignored.

- *Scalability* – The time-consuming, costly, and labour-intensive aspect of incorporating self-reported scales as means of assessing soft skills limits the frequency and coverage of these approaches. Self-reported measures lack scalability and cannot be deployed to measure skill development among a wider audience. Similarly, the administration of survey-type questionnaires does not guarantee total participation (Pongpaichet et al., 2022). Furthermore, monitoring the progression of these complex skills over time is vital and critical for enhancing learning outcomes (Dawson & Siemens, 2014). However, administering the same questionnaire repeatedly to measure growth can result in burnout (Sutherland et al., 2013).
- *Pre/post-test* – While adopting a pre- and post-test approach to measure skills development has been a prominent approach, such techniques do not account for the learning taking place during the study period. Pre-post assessment models developed for measuring leadership skills are usually deployed before and after learning content delivery (Amagoh, 2009) and provide snapshots of learning overtime. As such, they cannot capture the learning progression of the learners through the different stages of skills development and how their learning is associated with the development.
- *Active assessment* – Learners are required to participate in assessment questionnaires and surveys during the study period; thus, interfering obtrusively with their learning processes. As such, there is a need to adopt unobtrusive methods to quietly assess soft skills and allow instructors to monitor learners' development and growth without interrupting the study flow (Pongpaichet et al., 2022).
- *Analytical techniques* – Traditional measurement models used in the field of psychometrics and learning assessments do not utilize the educational data generated by online learning platforms to the full extent. Measurement models used by psychometricians usually rely on fixed-item responses by participants to measure learners' knowledge about subject content without necessarily considering the learning strategies adopted by participants while solving tasks. Traditional assessment techniques developed for the analysis of test responses cannot be applied to educational trace data. As such, the existing approaches do not consider the learning process and what learners do and only focus on the learning outcome. In this regard, there is a need to link the assessment of student learning outcomes and their learning behaviour and strategies to effectively identify the overall progress. In contrast, the fields of Educational Data Mining (EDM) and Learning Analytics (LA) have utilized trace data to provide unobtrusive means of assessing the learning strategies adopted by learners within MOOCs; thus, contributing to a richer understanding of the complex behaviour associated with student learning (Dawson & Siemens, 2014) without interfering with the dynamic learning process. Also, the use of trace data collected from various educational technologies eliminates biases generated from self-reported measures and the effort of administering additional instruments to collect data (Gray & Bergner, 2022). However, the statistical relations found in these LA studies only demonstrate that these patterns are unlikely random but can be inconsequential in judging an individual's learning (Milligan, 2020). The probabilistic dependency of

the observed variables on the targeted latent skill/learning objectives is often missing within LA (Mislevy et al., 2012). Although there are individual limitations in both these educational assessment fields, several studies have adopted multi-disciplinary techniques that draw on the strengths of one another for providing a holistic assessment of soft skills (Milligan & Griffin, 2016).

While the persistence of these challenges limits the measurement of soft skills, more reliable measures can be achieved through the careful consideration of unobtrusive approaches that go beyond self-reported scores. Therefore, by building on some of the earlier works in LA and implementing advanced analytical methods intersecting measurement models, we propose more scalable and unobtrusive means of assessing soft skills.

3 Case Study

3.1 Study Context

This chapter extracts data from an online professional learning program to develop leadership capabilities among the employees of a large global US corporation. The participants of the program were full-time working professionals and were mainly from the engineering and management domains, with varying professional backgrounds ranging from fresh graduates to individuals with over 15 years of experience. Delivered as a part of workplace training, this program was hosted in the Open_edX platform and was made available for free to all its employees.

This program consisted of a series of four Massive Open Online Courses (MOOCs) covering different aspects of leadership development. The first course of the program was scheduled for 4 weeks, while the remaining three courses were 3 weeks long each, delivered consecutively with a week-long break between two MOOCs. Also, these were asynchronous MOOCs that were designed to deliver several leadership learning objectives through recorded learning videos and related learning modules. Additionally, the MOOCs also included various formative assessments such as quizzes and self-reflection questions and summative essay assessments on several leadership concepts. All these assessments contributed to the certification grade for each MOOC. The second course of the program, however, followed a different instructional design and was left out of the analyses. The three components (MOOCs) of leadership investigated were – *understanding organizational strategy and capability, leading change in organizations*, and *discovering and implementing individual leadership strengths*.

In this chapter, we are particularly interested in the assessment of the self-reflection answers and use it as a proxy to comprehend and quantify leadership development (Helyer, 2015). The other kinds of formative assessments, such as the polls and the multiple-choice questions, allowed multiple attempts and prompted learners with hints. As such, the answers to these formative assessments may not adequately measure the development of the skills (Barthakur et al., 2022a).

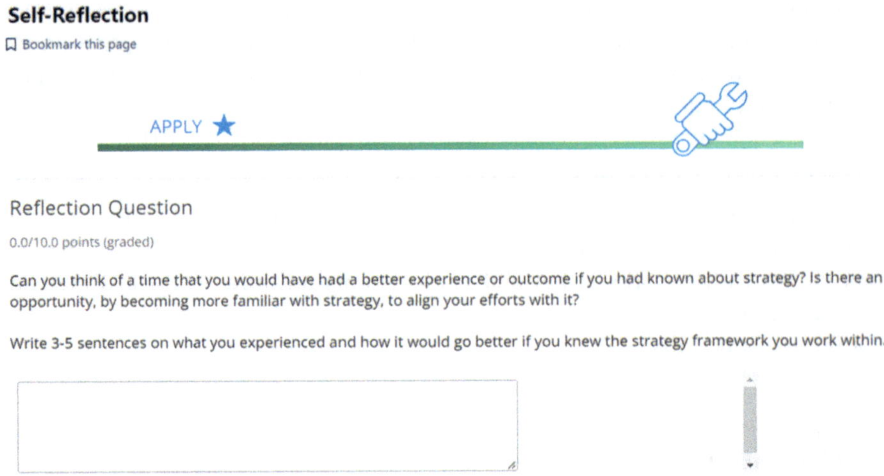

Fig. 4.1 An example of a self-reflection assessment question

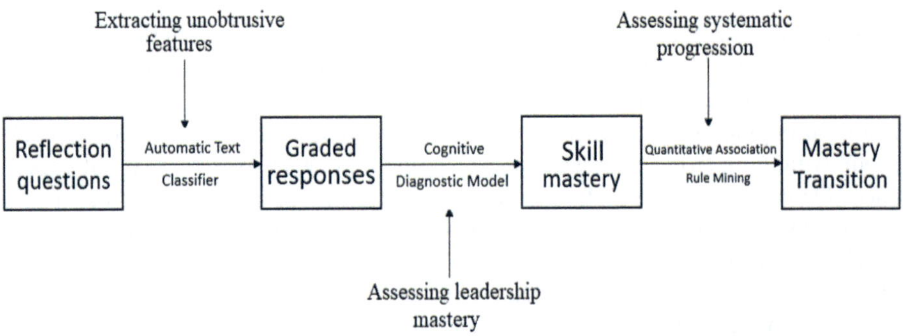

Fig. 4.2 The methodological pipeline used in the outlined case study. (Adopted from Barthakur et al., 2022a)

The self-reflection questions used within these MOOCs were content-specific in the sense that learners were encouraged to reflect on their learnings and experiences from leadership perspectives (Fig. 4.1). While discourse analysis and, more particularly, automated assessment of reflection has been studied by LA researchers for some time (Buckingham Shum et al., 2017; Jung & Wise, 2020; Ullmann, 2019), there is limited research on the assessment of reflection depth by adult learners (Barthakur et al., 2022b). Furthermore, although literature shows the role of reflection in skill development (Densten & Gray, 2001; Helyer, 2015; Wu & Crocco, 2019), there is a dearth of studies focusing on using reflection assessments as an unobtrusive means for evaluating soft skills.

The overview of the methodological pipeline adopted in the case study is provided in Fig. 4.2. In this, we adopt a blended methodology intersecting LA techniques and psychometric measurement models (Drachsler & Goldhammer, 2020).

This novel methodology draws on the strengths of different disciplines and mitigates the challenges listed in Sect. 2.3, thus, providing a means for unobtrusively measuring leadership skills.

3.2 *Extracting Unobtrusive Measures*

As we previously discussed, assessments evaluating soft skills are often actively administered, requiring learners to respond explicitly to questionnaires or other self-reported scales (Pongpaichet et al., 2022). In contrast, the analysis of the text responses to self-reflection questions can provide an unobtrusive means of evaluating leadership growth and acquisition, provided that researchers can extract features indicative of leadership skill mastery.

In one of our recent works (Barthakur et al., 2022b), we outlined a methodology to extract unobtrusive features from written reflective artefacts by implementing quantitative content analysis (Krippendorff, 2003) and developed an automated assessment system. The reflection responses varied in length and in the range of 17–393 words, with an average of 74 words utilized across each of the fifteen different questions. Extracting data from 771 out of the 861 learners who attempted all the reflection questions, the responses were categorized into four different hierarchical levels depending on the depth of reflection exhibited. These four levels were coded in accordance with a reflection framework developed by Kember et al. (2008) and are as follows – No-reflection, Understanding, Simple reflection and Critical reflection.

Two independent human coders manually graded a hundred answers each for the first four questions. When inter-rater reliability of 0.70 was achieved, the workload was equally divided to code the remaining answers to the same four questions. Using the manually coded responses as the training set and extracting several linguistic features from the written artefacts (such as Linguistic Word Count Inquiry, Coh-Metrix, n-grams, and readability index, among others), a machine learning classifier was trained to automatically analyse the answers to the remaining reflective assessments from the first MOOC. The performance of these models was judged based on their accuracy (closeness of the predicted values to the true values) and AUC ROC (Area Under the Curve – Receiver Operating Characteristics) values. While the model achieved a moderate accuracy of 0.66 and an AUC ROC of 0.88, the focus of the study was on establishing explainable insights rather than achieving higher accuracy through the implementation of AI black boxes (Dawson et al., 2019; Sartori & Theodorou, 2022).

Overall, the use of an automatic assessment approach provided the means for extracting unobtrusive measures of four reflection levels that were used to build models assessing the mastery of leadership learning objectives. Besides categorizing the responses into different levels based on the depth of reflection, the top twenty linguistic features predictive of the four levels were also analysed (Fig. 4.3). These

Feature	Description	SHAP	Self-Reflection Levels			
			No-Reflection	Understanding	Simple Reflection	Critical Reflection
WC	Word count of each answer	109.07	46.86 (22.91)	64.73 (22.58)	104.32 (33.82)	164.85 (52.27)
cm.WRDPRO	Word information: Pronoun indices	29.33	10.13 (5.06)	15.23 (5.98)	24.22 (9.26)	37.76 (14.63)
cm.WRDPRP1s	Word information: First person singular pronoun incidence	28.6	11.12 (5.73)	14.82 (5.61)	23.72 (7.82)	37.3 (11.14)
cm.LDVOCDa	Lexical Diversity: Lexical diversity, VOCD, all words	17.75	0.39 (0.41)	0.69 (0.28)	0.81 (0.05)	0.82 (0.03)
MD (POS Tagging)	Modal Verbs (e.g., can, could)	12.67	0.57 (0.95)	0.78 (1.02)	1.32 (1.27)	2.24 (1.85)
Product	Frequency of "Product" Unigram	11.58	0.18 (0.52)	0.57 (1.0)	0.46 (1.05)	0.7 (1.51)
Thing	Frequency of "Thing" Unigram	6.19	0.2 (0.55)	0.13 (0.45)	0.23 (0.64)	0.37 (0.88)
cm.DRPVAL	Syntactic pattern density: Agentless passive voice density incidence	5.59	11.96 (6.02)	16.82 (6.01)	26.54 (8.89)	41.32 (13.48)
NN (POS tagging)	Singular nouns (e.g., business, college)	5.16	7.86 (4.25)	11.4 (5.02)	18.0 (7.46)	28.23 (10.1)
Resource	Frequency of "Resource" Unigram	4.73	0.13 (0.37)	0.37 (0.71)	0.27 (0.7)	0.46 (1.03)
IN (POS tagging)	Preposition or subordinating conjunction	4.03	0.67 (0.86)	1.1 (1.05)	1.84 (1.45)	3.09 (2.01)
cm.CRFANPa	Referential Cohesion: Anaphor overlap, all sentences	3.61	0.25 (0.38)	0.37 (0.36)	0.41 (0.33)	0.43 (0.31)
Control	Frequency of "Control" Unigram	3.4	0.19 (0.6)	0.12 (0.54)	0.2 (0.71)	0.29 (0.87)
Would	Frequency of "Would" Unigram	3.18	0.38 (0.74)	0.51 (0.83)	0.89 (1.08)	1.52 (1.46)
Liwc.Dic	Dictionary words	2.98	88.67 (6.54)	88.48 (5.54)	87.52 (5.0)	88.1 (4.42)
cm.DRNEG	Syntactic pattern density: Negation density, incidence	2.91	8.59 (4.33)	11.3 (4.19)	18.17 (6.17)	28.34 (8.67)
difficult_words	Words not matching the Dale-Chall list of "familiar" words	2.44	10.21 (5.03)	13.93 (5.55)	21.81 (8.26)	33.17 (12.07)
Liwc.focuspresent	Time orientation: focus towards present	2.4	10.54 (5.1)	8.88 (4.57)	8.09 (4.07)	8.51 (3.67)
Liwc.money	Personal concerns: Money (e.g., audit, cash, owe)	2.05	0.77 (1.8)	1.2 (1.9)	1.07 (1.58)	0.88 (1.14)
Liwc.you	Personal pronouns: Second person (e.g., you, your)	1.85	0.79 (2.12)	0.72 (2.11)	0.25 (0.96)	0.34 (0.91)
Liwc.tentat	Cognitive processes: Tentative (e.g., maybe, perhaps)	1.82	2.27 (2.58)	2.3 (2.27)	2.15 (1.83)	2.39 (1.68)
Project	Frequency of "Project" Unigram	1.81	0.2 (0.49)	0.61 (0.99)	0.63 (1.23)	0.85 (1.52)
PRP (POS Tagging)	Personal pronouns	1.73	1.69 (1.6)	2.09 (1.93)	3.51 (2.29)	5.07 (3.3)
Liwc.insight	Cognitive Processes: Insight (e.g., think, know, consider)	1.67	4.16 (3.5)	3.47 (2.92)	3.7 (2.49)	3.54 (2.02)
Liwc.Ner	Negations (e.g., no, not, never)	1.64	0.49 (0.96)	0.71 (1.18)	1.45 (1.7)	2.32 (2.38)
Liwc.Adverb	Adverbs (e.g., really, very)	1.58	3.29 (3.01)	3.1 (2.51)	3.04 (1.93)	3.25 (1.6)
cm.SYNMEDpos	Syntactic Complexity: Minimal Edit Distance, part of speech	1.37	8.26 (3.67)	10.43 (3.48)	15.52 (4.57)	21.69 (5.31)
RBR (POS tagging)	Adverbs, comparative	1.2	1.52 (1.58)	1.89 (1.66)	3.04 (2.16)	5.21 (2.92)
cm.SYNMEDlem	Syntactic Complexity: Minimal Edit Distance, lemmas	1.19	4.85 (5.05)	5.71 (4.48)	5.73 (3.77)	7.45 (4.3)
People	Frequency of "People" Unigram	1.15	0.07 (0.28)	0.05 (0.24)	0.1 (0.38)	0.14 (0.43)

Fig. 4.3 Top twenty features summary and their association with the four reflection levels. (Adopted from Barthakur et al., 2022b)

features are ranked based on their SHAP score (a unified measure of feature importance), and the association with the four levels is also provided.

While some of the findings echo that of previous studies, such as higher word count being indicative of a higher level of reflective practice, several newer insights regarding reflection in relation to skill development were discovered. For instance, it was observed that learners tend to describe more about their present learning and professional experiences while developing skills compared to other learners in more traditional settings focusing on past events (Kovanovic et al., 2018). Another observation, captured through the readability index, includes the use of more complicated phrases in higher levels of reflective text, while learners engaging in shallow reflection are less expressive (with fewer word counts) and tend to rely on simple dictionary words. Also, the use of first-person and second-person (such as you and your) personal pronouns can help identify the depth of reflection exhibited by the learners. Such insights were previously unknown, are critical for comprehending (leadership) skill development, and have important practical implications (Barthakur et al., 2022b). Based on the reflection levels, such findings can also provide opportunities for supporting and scaffolding learners. Learners demonstrating shallow reflection can be provided real-time feedback and enhance their skill acquisition during the learning process.

3.3 Assessing Leadership Mastery

In Sect. 2.2, we highlighted the role of reflective practices in different educational contexts and, more particularly, in developing leadership skills. Extending the above methodology and the extracted unobtrusive measures of reflection on leadership concepts, another research project evaluated the mastery of leadership skills defined as learning objectives in MOOCs (Barthakur et al., 2022c). While the limitation of lack of probabilistic dependencies between the observed variables and the latent learning constructs is often discussed in LA studies (Milligan, 2020), in this second component of the case study, mastery of (five) latent leadership objectives was calculated using the ordinal four-level graded reflections and a probabilistic relationship was established.

In this example, we divided the analysis into two steps – providing an assessment of the mastery of the individual skills based on the reflection grades and finding clusters of students based on their mastery of all skills in the course. First, a measurement model (cognitive diagnostic model, CDM; Lee & Sawaki, 2009; Rupp et al., 2010) was implemented using the four-level graded reflection responses as the input. CDMs are person-centred models that are used when empirical information about latent skills and attributes is sought (Rupp & Templin, 2008). CDMs in this case study were used to calculate probabilities of (latent) leadership skill mastery for all the learners based on their written artefacts. These models provide information about the extent of mastery of these latent skills, in the range of zero to one. A probability closer to one demonstrates higher mastery, while probabilities on the other side of the spectrum closer to zero indicate lower levels of mastery. Out of the several types of CDM models, a generalized model was chosen given its generalizability and relaxed nature (devoid of any strong conjunctive or disconjunctive assumptions). Usually, most CDM models are constrained and require fulfilling several assumptions as compared to the generalized CDM. Furthermore, due to the ordinal nature of the graded reflection data, a sequential CDM model was used for this analysis (Barthakur et al., 2022c). In the second and final step of the analysis, the learners were then categorized into different groups using a clustering algorithm to develop a holistic understanding of skill mastery at the cohort level in the entire MOOC (Fig. 4.2).

As mentioned earlier, these leadership skills within each MOOC are defined as learning objectives. The probabilities calculated by the CDM models represent the extent of mastery for the various individual learning objectives; probabilities closer to one generally indicate higher mastery. Such results provide diagnostic information about their acquisition and mastery across several leadership components inferred from the reflective responses. Additionally, based on the probabilities of skill mastery, the clustering algorithm identified four distinct learning profiles (Fig. 4.4). These profiles were labelled depending on the average learning objective mastery. It was observed that the latter learning objectives of the MOOC had a higher mastery rate while the lowest mastery across the first learning objective across all the profiles. We supported these findings through leadership theory and propose that the learners were building on the contents associated with the earlier learning objectives to exhibit superior mastery as the course progressed. The gradual shift in the learning objective mastery highlight "the effect of the design and sequencing of individual learning objectives on content mastery" (Barthakur et al., 2022a, b, c, p. 17).

This component of the case study extends the discussion of understanding and assessing reflection answers by operationalizing meaningful latent learning objectives. Based on the depth of reflection exhibited by the learners in the written artefacts, a measurement model was implemented to calculate the mastery of various leadership learning objectives. An important implication of such an analysis is that it shifts the focus from evaluating learners' cognitive knowledge based on their final course grades to the assessments of skill mastery. It also advances the discussion of learner profiling by categorizing learners based on the evidenced learning objective mastery, which was previously done using behavioural engagement data and final course grades. Moreover, while the study in Sect. 3.2 can be used for scaffolding learners with individual reflection questions, the findings from this component allow instructors and other stakeholders to provide pedagogical interventions depending on learners' mastery of learning objectives and support learners with specific content within the course. From the methodological standpoint, this work provides a novel blended methodology approach marrying learning analytics and measurement theory models to measure soft skills. The two studies discussed above combined provide an overview of the novel implementation of unobtrusive approaches to soft skills assessment in digital learning settings, opening avenues to

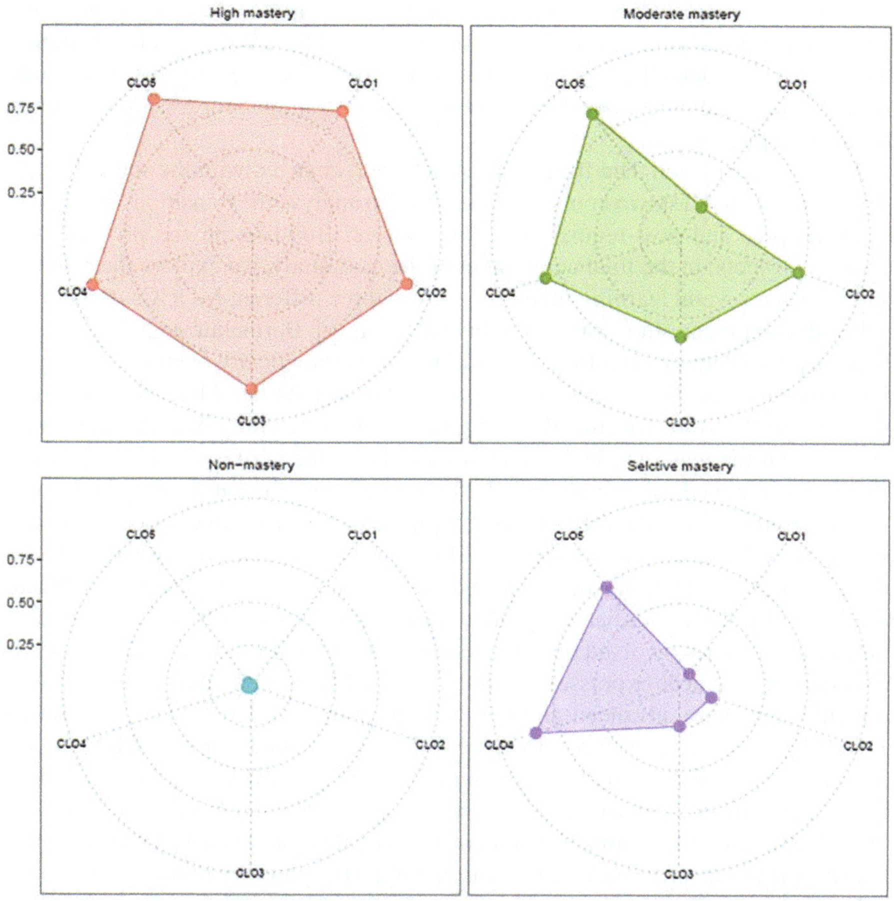

Fig. 4.4 Learner profiles are based on average learning objective mastery. (Adopted from Barthakur et al., 2022c)

extend the methodology to measure their longitudinal progress and growth over time and across several courses.

3.4 Assessing Systematic Progression

Effective and thoughtful sequencing of courses and learning contents are critical for allowing learners to successfully navigate and acquire knowledge while traversing through a study program (Dawson & Hubball, 2014). Study programs with either flexible or restricted pathways, when effectively structured, can often reduce learners' cognitive overload and enhance academic performance (Barthakur et al., 2022a). While the effectiveness of the courses and learning content sequencing are primarily evaluated using introspective peer-review approaches, these measures have several drawbacks, as suggested in Sect. 2.2. However, the introduction of online micro-credential programs, such as the one discussed in this case study, opens avenues to collect trace data to evaluate the systematic progression of learners across multiple courses.

This work extracted data from 771 learners who engaged with the self-reflection assessments in at least two courses, allowing us to analyse the transitions and understand the pre- and post-requisites of the courses. Building on the previous two research projects, in the final component of the case study, we explore the relationship between several learning objectives across three different MOOCs of the leadership development study program. In the works of Barthakur et al. (2022a), a three-step blended methodology was outlined to automatically evaluate these relationships based on the empirical assessment data across the whole MOOC study program (*ibid*). More specifically, using the machine learning classifier described above to automatically grade the reflective artefacts, the mastery of learning objectives was calculated using multiple CDM models across the entire study program.

In the third stage of the methodological pipeline, a Quantitative Association Rule Mining (QARM; Salleb-Aouissi et al., 2007) was implemented to investigate learners' transitions in learning objective mastery when traversing across the MOOC program. QARM is similar to general association rules with the exception of numerical attributes involved on either side of the rule. For instance, while a general association rule can be expressed as {Butter} → {Milk, Flour}, quantitative association rules are more advanced and can be expressed as {2 Butter} → {3 Milk, 1 Flour}. In this current example, the probabilities calculated from the CDMs in the previous step were converted into three ordinal levels – low mastery (probabilities below 0.60), medium mastery (between 0.60 and 0.80) and high mastery (above 0.80). In doing so, the ordinal levels serve as adequate input to the QARM algorithm and support the identification of the learning objective mastery relationship.

From the first part of the analysis, it was observed that the learners exhibited varying probabilities of learning objective mastery across the three courses. However, the unique contribution of this example is the analysis of mastery transition and understanding how prerequisites in a course affect the mastery of content in the subsequent courses of a study program. Barthakur et al. (2022a) traced some of these findings in various seminal (Quinn, 1988) and modern (Corbett, 2021; Corbett & Spinello, 2020) leadership theories and frameworks. Interpreting the mastery transitions (Fig. 4.5), it was observed that higher mastery across the five leadership objectives in the first MOOC resulted in higher mastery across the first (3.1) and third (3.3) objectives of the third MOOC. On the contrary, failing to demonstrate high command over the learning objectives of the first MOOC can significantly affect the mastery of the last learning objective (3.4) of the third MOOC. Similar observations of low mastery can be made in the second objective (4.2) of the fourth MOOC when failing to master the leadership objectives of the third MOOC on *leading change in organizations*. Such a relationship echoes Quinn's (1988) theories of the role of effective leadership in facilitating change to enhance organizational performance.

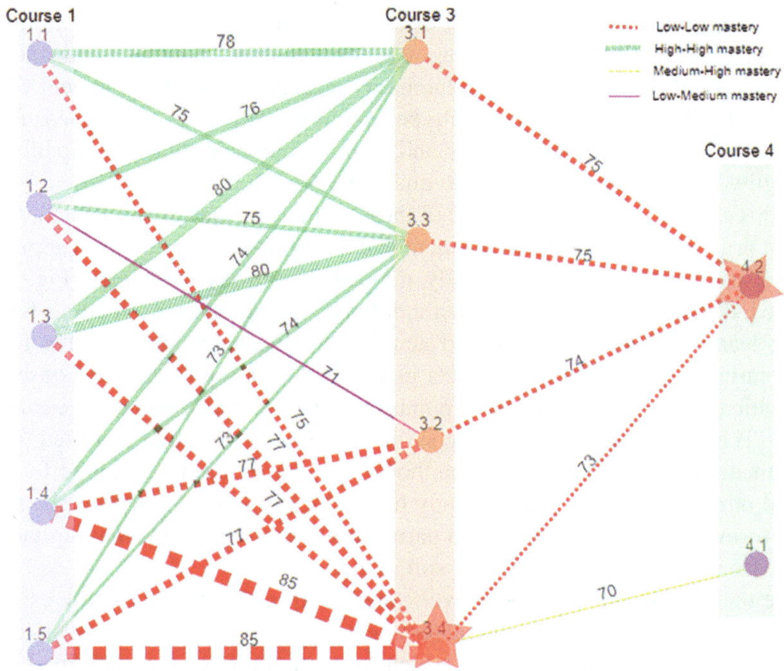

Fig. 4.5 Transitions in leadership objective mastery across a study program. (Adopted from Barthakur et al., 2022a)

Using an evidence-based approach, this work contributes to our understanding of learners' skill mastery within and across multiple MOOCs in the study program and how they transition over time. Such findings can provide instructors with information about students' learning which in turn can be used to provide pedagogically informed decisions. For instance, learners exhibiting lower mastery in the first course can be supported with additional resources to successfully complete the final objective (3.4) of the third MOOC. Similarly, these findings can be used for gathering diagnostic fine-grained information about the mastery of individual learning objectives that go beyond analysing learners' success based on final course grades. Finally, from the perspective of the course designers, instructors, and researchers, this will allow for investigating the ordering of learning objectives and courses to reduce cognitive overload and enhance student learning experiences.

4 Conclusion

The importance of soft skills in the modern workforce has been extensively discussed in the last few decades. Several frameworks have been conceptualized to comprehend and promote the development of these complex skills (Casner-Lotto &

Barrington, 2006). While significant efforts were made in terms of promoting *soft skills*, there were significant challenges in the way these competencies and skills were measured. Most previous research has measured skill development through the use of subjective questionnaire-type measures. However, these measures are often associated with several biases and cannot guarantee total participation. The scalability of such approaches is also questionable.

In this chapter, we illustrate some of these challenges that are associated with the current practices of soft skill assessment and a need for evidence-based approaches for measuring soft skills. A case study, divided into three components (Fig. 4.2), is outlined, describing a data-driven methodology using unobtrusive features for measuring leadership skill mastery and acquisition. Extracting unobtrusive features from learners' self-reflection artefacts in a MOOC study program, responses were automatically graded, and leadership mastery was calculated using a measurement model. The interdependencies of the skills' mastery were further analysed to study the transitions over time. Such a methodology can be easily extended to extract several other unobtrusive features (any hierarchically graded assessments, such as in the form of correct, incorrect, and partially correct responses) from digital learning environments to assess different soft skills.

The underlying premise of the work presented in this chapter revolves around advancing research related to the assessment of complex soft skills. This chapter outlines three studies that illustrate a blended methodology by combining learning analytics and psychometrics to measure leadership skills by collecting digital assessment data from a professional development MOOC program. These unobtrusive approaches to data extraction provide the opportunity to passively measure the development of skills without interfering with the learning processes. The assessment models discussed are fully automatic and thus have the potential to be implemented at scale. Finally, the generalizability of the approach allows the assessment of other skills in varying contexts.

References

Amagoh, F. (2009). Leadership development and leadership effectiveness. *Management Decision, 47*(6), 989–999. https://doi.org/10.1108/00251740910966695

Bachman, J. G., & O'Malley, P. M. (1984). Yea-saying, Nay-saying, and going to extremes: Black-White differences in response styles. *Public Opinion Quarterly, 48*(2), 491–509. https://doi.org/10.1086/268845

Barthakur, A., Joksimovic, S., Kovanovic, V., Corbett, F. C., Richey, M., & Pardo, A. (2022a). Assessing the sequencing of learning objectives in a study program using evidence-based practice. *Assessment & Evaluation in Higher Education*, 1–15. https://doi.org/10.1080/02602938.2022.2064971

Barthakur, A., Joksimovic, S., Kovanovic, V., Ferreira Mello, R., Taylor, M., Richey, M., & Pardo, A. (2022b). Understanding depth of reflective writing in workplace learning assessments using machine learning classification. *IEEE Transactions on Learning Technologies*, 1. https://doi.org/10.1109/TLT.2022.3162546

Barthakur, A., Kovanovic, V., Joksimovic, S., Zhang, Z., Richey, M., & Pardo, A. (2022c). Measuring leadership development in workplace learning using automated assessments: Learning analytics and measurement theory approach. *British Journal of Educational Technology*. https://doi.org/10.1111/bjet.13218

Bergner, Y. (2017). Measurement and its uses in learning analytics. In C. Lang, G. Siemens, A. Wise, & D. Gasevic (Eds.), *Handbook of learning analytics* (1st ed., pp. 35–48). Society for Learning Analytics Research (SoLAR). https://doi.org/10.18608/hla17.003

Buckingham Shum, S., Sándor, Á., Goldsmith, R., Bass, R., & McWilliams, M. (2017). Towards reflective writing analytics: Rationale, methodology and preliminary results. *Journal of Learning Analytics, 4*(1), 10.18608/jla.2017.41.5.

Burke, V., & Collins, D. (2005). Optimising the effects of leadership development programmes: A framework for analysing the learning and transfer of leadership skills. *Management Decision, 43*(7/8), 975–987. https://doi.org/10.1108/00251740510609974

Care, E., Griffin, P., & Wilson, M. (Eds.). (2018). *Assessment and teaching of 21st century skills: Research and applications*. Springer. https://doi.org/10.1007/978-3-319-65368-6

Casner-Lotto, J., & Barrington, L. (2006). Are they really ready to work? Employers' perspectives on the basic knowledge and applied skills of new entrants to the 21st century U.S. workforce. In *Partnership for 21st century skills*. Partnership for 21st Century Skills. https://eric.ed.gov/?id=ED519465

Chauhan, A. (2014). *Massive open online courses (MOOCS): Emerging trends in assessment and accreditation* (p. 12).

Corbett, F. (2021). *Emergence of the connectivist leadership paradigm: A grounded theory study in the Asia region*. Theses and dissertations. https://digitalcommons.pepperdine.edu/etd/1194

Corbett, F., & Spinello, E. (2020). Connectivism and leadership: Harnessing a learning theory for the digital age to redefine leadership in the twenty-first century. *Heliyon, 6*(1), e03250. https://doi.org/10.1016/j.heliyon.2020.e03250

Dawson, S., & Hubball, H. (2014). Curriculum analytics: Application of social network analysis for improving strategic curriculum decision-making in a research – Intensive university. *Learning Inquiry, 2*(2), 59–74.

Dawson, S., & Siemens, G. (2014). Analytics to literacies: The development of a learning analytics framework for multiliteracies assessment. *International Review of Research in Open and Distance Learning, 15*, 284–305. https://doi.org/10.19173/irrodl.v15i4.1878

Dawson, S., Joksimovic, S., Poquet, O., & Siemens, G. (2019). Increasing the impact of learning analytics. In *Proceedings of the 9th international conference on learning analytics & knowledge* (pp. 446–455). https://doi.org/10.1145/3303772.3303784

Densten, I., & Gray, J. (2001). *Leadership development and reflection: What is the connection?* (p. 15). http://lst-iiep.iiep-unesco.org/cgi-bin/wwwi32.exe/[in=epidoc1.in]/?T2000=013168/(100). https://doi.org/10.1108/09513540110384466

Drachsler, H., & Goldhammer, F. (2020). Learning analytics and eAssessment – Towards computational psychometrics by combining psychometrics with learning analytics. In D. Burgos (Ed.), *Radical solutions and learning analytics: Personalised learning and teaching through big data* (pp. 67–80). Springer. https://doi.org/10.1007/978-981-15-4526-9_5

Ebrahimi, M. S., & Azmi, M. N. (2015). New approach to leadership skills development (developing a model and measure). *Journal of Management Development, 34*(7), 821–853. https://doi.org/10.1108/JMD-03-2013-0046

Ginda, M., Richey, M. C., Cousino, M., & Börner, K. (2019). Visualizing learner engagement, performance, and trajectories to evaluate and optimize online course design. *PLoS ONE, 14*(5), e0215964. https://doi.org/10.1371/journal.pone.0215964

Gray, G., & Bergner, Y. (2022). A practitioner's guide to measurement in learning analytics – Decisions, opportunities, and challenges. In *Handbook of learning analytics* (2nd ed., pp. 20–28).

Haste, H. (2001). Ambiguity, autonomy and agency: Psychological challenges to new competence. *Defining and Selecting Key Competencies*, 93–120.

Helyer, R. (2015). Learning through reflection: The critical role of reflection in work-based learning (WBL). *Journal of Work-Applied Management, 7*(1), 15–27. https://doi.org/10.1108/JWAM-10-2015-003

Jenkins, H., Clinton, K., Purushotma, R., Robison, A. J., & Weigel, M. (2006). *Confronting the challenges of participatory culture: Media education for the 21st century*. MacArthur Foundation.

Joksimovic, S., Kovanovic, V., & Dawson, S. (2019). The journey of learning analytics. *HERDSA Review of Higher Education, 6*, 37–63.

Joksimovic, S., Siemens, G., Wang, Y. E., San Pedro, M. O. Z., & Way, J. (2020). Editorial: Beyond cognitive ability. *Journal of Learning Analytics, 7*(1), 1–4. https://doi.org/10.18608/jla.2020.71.1

Jung, Y., & Wise, A. F. (2020). How and how well do students reflect? Multi-dimensional automated reflection assessment in health professions education. In *Proceedings of the tenth international conference on learning analytics & knowledge* (pp. 595–604). https://doi.org/10.1145/3375462.3375528

Kember, D., McKay, J., Sinclair, K., & Wong, F. K. Y. (2008). A four-category scheme for coding and assessing the level of reflection in written work. *Assessment & Evaluation in Higher Education, 33*(4), 369–379. https://doi.org/10.1080/02602930701293355

Kovanović, V., Joksimović, S., Mirriahi, N., Blaine, E., Gašević, D., Siemens, G., & Dawson, S. (2018). Understand students' self-reflections through learning analytics. In *Proceedings of the 8th international conference on learning analytics and knowledge* (pp. 389–398). https://doi.org/10.1145/3170358.3170374.

Krippendorff, K. (2003). *Content analysis: An introduction to its methodology* (p. 8). Sage.

Kyllonen, P. C. (2012). Measurement of 21st century skills within the common core state standards. In *Invitational research symposium on technology enhanced assessments* (p. 24).

Lai, E. R., & Viering, M. (2012). *Assessing 21st century skills: Integrating research findings*. Pearson.

Lee, Y.-W., & Sawaki, Y. (2009). Cognitive diagnosis approaches to language assessment: An overview. *Language Assessment Quarterly, 6*(3), 172–189. https://doi.org/10.1080/15434300902985108

Martin, C. K., Nacu, D., & Pinkard, N. (2016). Revealing opportunities for 21st century learning: An approach to interpreting user trace log data. *Journal of Learning Analytics, 3*(2), 37–87.

Milligan, S. (2020). Standards for developing assessments of learning using process data. In M. Bearman, P. Dawson, R. Ajjawi, J. Tai, & D. Boud (Eds.), *Re-imagining university assessment in a digital world* (Vol. 7, pp. 179–192). Springer. https://doi.org/10.1007/978-3-030-41956-1_13

Milligan, S. K., & Griffin, P. (2016). Understanding learning and learning design in MOOCs: A measurement-based interpretation. *Journal of Learning Analytics, 3*(2), 88–115. https://doi.org/10.18608/jla.2016.32.5

Mislevy, R. J., Behrens, J. T., Dicerbo, K. E., & Levy, R. (2012). Design and discovery in educational assessment: Evidence-centered design, psychometrics, and educational data mining. *Journal of Educational Data Mining, 4*(1), 11–48.

Nederhof, A. J. (1985). Methods of coping with social desirability bias: A review. *European Journal of Social Psychology, 15*(3), 263–280. https://doi.org/10.1002/ejsp.2420150303

Pellegrino, J. W. (2017). *Teaching, learning and assessing 21st century skills* (pp. 223–251). https://doi.org/10.1787/9789264270695-12-en

Pongpaichet, S., Nirunwiroj, K., & Tuarob, S. (2022). Automatic assessment and identification of leadership in college students. *IEEE Access*, 1. https://doi.org/10.1109/ACCESS.2022.3193935

Quinn, R. E. (1988). *Beyond rational management: Mastering the paradoxes and competing demands of high performance* (pp. xxii, 199). Jossey-Bass.

Rios, J. A., Ling, G., Pugh, R., Becker, D., & Bacall, A. (2020). Identifying critical 21st-century skills for workplace success: A content analysis of job advertisements. *Educational Researcher, 49*(2), 80–89. https://doi.org/10.3102/0013189X19890600

Rohs, F. R., & Langone, C. A. (1997). Increased accuracy in measuring leadership impacts. *Journal of Leadership Studies, 4*(1), 150–158. https://doi.org/10.1177/107179199700400113

Rupp, A. A., & Templin, J. L. (2008). Unique characteristics of diagnostic classification models: A comprehensive review of the current state-of-the-art. *Measurement: Interdisciplinary Research and Perspectives, 6*(4), 219–262. https://doi.org/10.1080/15366360802490866

Rupp, A. A., Templin, J., & Henson, R. A. (2010). *Diagnostic measurement: Theory, methods, and applications*. Guilford Press.

Salleb-Aouissi, A., Vrain, C., & Nortel, C. (2007). *QuantMiner: A genetic algorithm for mining quantitative association rules* (pp. 1035–1040).

Sartori, L., & Theodorou, A. (2022). A sociotechnical perspective for the future of AI: Narratives, inequalities, and human control. *Ethics and Information Technology, 24*(1), 4. https://doi.org/10.1007/s10676-022-09624-3

Shum, S. B., & Crick, R. D. (2016). Learning analytics for 21st century competencies. *Journal of Learning Analytics, 3*(2), 6–21. https://doi.org/10.18608/jla.2016.32.2

Sondergeld, T. A., & Johnson, C. C. (2019). Development and validation of a 21st century skills assessment: Using an iterative multimethod approach. *School Science and Mathematics, 119*(6), 312–326. https://doi.org/10.1111/ssm.12355

Sutherland, M. A., Amar, A. F., & Laughon, K. (2013). Who sends the email? Using electronic surveys in violence research. *The Western Journal of Emergency Medicine, 14*(4), 363–369. https://doi.org/10.5811/westjem.2013.2.15676

Ullmann, T. D. (2019). Automated analysis of reflection in writing: Validating machine learning approaches. *International Journal of Artificial Intelligence in Education, 29*(2), 217–257. https://doi.org/10.1007/s40593-019-00174-2

Vockley, M. (2007). Maximizing the impact: The pivotal role of technology in a 21st century education system. In *Partnership for 21st century skills*. Partnership for 21st Century Skills. https://eric.ed.gov/?id=ED519463

Wu, Y., & Crocco, O. (2019). Critical reflection in leadership development. *Industrial and Commercial Training, 51*(7/8), 409–420. https://doi.org/10.1108/ICT-03-2019-0022

[illegible]

Chapter 5
Reconfiguring Measures of Motivational Constructs Using State-Revealing Trace Data

Heeryung Choi, Philip H. Winne, and Christopher Brooks

Abstract This chapter examines opportunities afforded by trace data to capture dynamically changing latent states and trajectories spanning states in self-regulated learning (SRL). We catalog and analyze major challenges in temporally investigating SRL constructs related to a prominent motivational factor, achievement goals. The dynamics of potentially frequent state changes throughout a learning session and across sessions are poorly reflected by self-report survey items typically administered before and after a session or, less informatively, at the beginning of an academic term. Trace data, carefully operationalized, offer substantial benefits compensating for shortcomings of comparatively static survey data. We summarize three recent studies addressing these challenges and characterize learning analytics designed to promote SRL and motivation formed from unobtrusive traces. This approach provides a practical and continuously updatable account of SRL constructs, varying dynamically within and across study sessions. We conclude by proposing a research agenda for learning analytics focusing on guiding and supporting SRL.

Keywords Trace data · Motivations · Achievement goals · Self-regulated learning · Dynamic SRL constructs

H. Choi (✉)
Massachusetts Institute of Technology, Cambridge, MA, USA
e-mail: heeryung@mit.edu

P. H. Winne
Simon Fraser University, Burnaby, BC, Canada
e-mail: winne@sfu.ca

C. Brooks
University of Michigan, Ann Arbor, MI, USA
e-mail: brooksch@umich.edu

V. Kovanovic et al. (eds.), *Unobtrusive Observations of Learning in Digital Environments*, Advances in Analytics for Learning and Teaching,
https://doi.org/10.1007/978-3-031-30992-2_5

1 Introduction: Self-Regulated Learning

Improvements in educational technologies have allowed researchers to integrate more unobtrusive trace data into their studies. Trace data are clickstream or log records designed to represent a specified theoretical construct revealed as learners operate on information while learning (Winne, 2020a, b). Trace data are particularly useful in measuring and researching dynamic properties of self-regulated learning (SRL).

Winne models how dynamic SRL states arise using the COPES model: Conditions, Operations, Products, Evaluations, and Standards (Winne, 1997, 2022). Self-regulating learners first identify internal and external *conditions* they perceive can affect tasks. Based on their understanding of those conditions, learners choose and carry out metacognitive and cognitive *operations*, generating *products* as a result. Learners then *evaluate* those products, including experiences arising from operations, gauging their properties using *standards*. For example, suppose a learner is preparing for an upcoming quiz in an Earth Science class. First, the learner considers factors such as knowledge about related topics, effort likely required, and incentives for earning a high grade. They recall difficulty listing the names of planets in the solar system and forecast if it is important to remember those to receive a satisfactory grade. Based on this understanding, they design a mnemonic device to assemble the names of planets in the solar system in serial order from the sun outward. After applying the mnemonic, they evaluate its utility using standards such as confidence they will be able to recall all the planets' names in the correct order and effort to encode this information.

SRL is recursive. While working on a given task, a learner could operationalize several COPES learning events to unfold SRL across the learning session and beyond. For example, after a cycle of SRL ends in the evaluation state, a learner might be highly satisfied with their product, such as the invented mnemonic device. This evaluation result could affect an upcoming SRL cycle; motivation might increase since they predict they could easily apply this same tactic to other cases – a high efficacy expectation – and review all the materials more quickly than they expected. Both have high incentive. That is, SRL is a dynamic progression of COPES learning events emerging and contingently unfolding throughout a task, a semester, or an academic year.

2 Dynamic Nature of Motivation

2.1 How to Capture Motivation

Motivation is an internal cognitive state that provides reasons for choices learners make about behavior (Kleinginna & Kleinginna, 1981; Winne & Marzouk, 2019). Motivation is often measured by asking learners to rate a motivational construct, such as achievement goals, or by time spent on tasks (Ames & Archer, 1988; Ames, 1992; Elliott & Dweck, 1988; Masgoret & Gardner, 2003). In the COPES model, motivation is a *condition* that contributes to (1) learners' plans, (2) choices about how to approach a task, and (3) forecasts about how to adapt to *operations* to improve work as tasks unfold. Motivation also plays an important role in the *evaluation* facet of COPES; motivation provides reasons for selecting *standards* to judge incentives associated with *products*.

Achievement goal theory generally explains goals using two dimensions: (1) mastery-performance and (2) approach-avoidance. The mastery-performance dimension differentiates the product learners pursue. It contrasts internal standards, such as joy and satisfaction for learning (i.e., mastery), vs. external standards, such as letter grades or ranking and performance with respect to peers. The approach-avoidance dimension contrasts whether learners: (1) seek to acquire desired stimuli (approach) or evade undesired stimuli (avoidance) (Ames, 1992; Nicholls, 1984).

As with other SRL constructs, motivation for achievement goals can dynamically change throughout a task and between tasks. For example, Muis and Edwards (2009) investigated goal changes between similar and different tasks. In both circumstances, they found evidence for both *goal switching*, replacing one goal with another, and *goal intensification*, increasing one's endorsement of an initial goal. They also found mastery-approach goals and performance-avoidance goals were less stable than performance-approach goals, a finding aligned to a previous study (Fryer & Elliot, 2007). Tuominen-Soini et al.'s (2011) studies also showed the dynamic nature of motivation, and they detected changes in Finnish students' achievement goals both between and within a school year. Using latent profile analysis of survey data to develop individual learners' motivational profiles, approximately 35% of students modulated their motivational profiles to reflect similar goal profiles while 5% of students completely changed their goals.

Considering the recursive nature of SRL, goal changes should be expected as learners traverse states in their work. Learners' initial goals may be formed using incomplete information about *conditions*, such as task difficulty. After some time, learners may update goals if *products* generated based on their incomplete understanding of conditions lead to an unsatisfactory evaluation relative to *standards*. In Fryer and Elliot's work (2007), substantial goal changes were more frequent after an initial task than after subsequent tasks, which showed how learners acquired more information from the initial encounter with a task and adjust their goals accordingly in the subsequent tasks.

When goals change, other COPES states may change accordingly, as learners may deem it useful to revise *operations*, hence affecting *products*. New *standards* for *evaluations* may also be adopted. For example, after trying a new strategy to solve the previously attempted math problems and evaluating the new strategy as successful, a learner may perceive greater efficacy for problem-solving and choose to attempt slightly more challenging "extra points" exercises. This goal might change again depending on the pace of work on those more challenging problems, with concomitant changes in self-efficacy depending on whether pace is evaluated as "fast" or "slow."

To more fully understand why motivation changes, and to predict more accurately if and how it will change, it is important to develop fuller accounts of the contexts in which change is observed. Without such contextual information, we suggest it will remain difficult to understand and assess learners' goal changes and design potential improvements to learning experiences.

2.2 A Role for Trace Data in Motivational Studies

Collecting contextual information about motivation and its roles in dynamic SRL is likely to be more informative and authentic if researchers adopt methods to gather fine-grained and unobtrusive data. Log and clickstream data, not yet common in motivational research, offer potential to sharpen research in two ways. First, if unobtrusively generated, such data can be gathered across the timeline of tasks with minimal to no interference. This affords detecting goal changes as they materialize in context. Second, because more detailed information can be captured about external conditions both preceding and at the point of change in motivational states, trace data set a stage for theorizing more productively about how to support learners' motivation while at the same time developing fuller pictures about how SRL relates to developing achievement outcomes. In this effort, it is important to engineer data gathering methods that minimize intrusions on and distortions to learners' authentic learning experiences.

In contrast to the potential benefits of online trace data to reflect changing conditions and motivational dynamics, motivation has been mostly measured using self-report measures – surveys and questionnaires – which often are not sufficiently fine-grained and task-specific (Winne, 2020a, b). The scope of that methodology is broad, ranging across studies and domains from sports to psychology to a residential mathematics course for K-12 learners to distance learning for adult learners (Elliot & Murayama, 2008; Jang & Liu, 2012; Luo et al., 2011; Remedios & Richardson, 2013; Wolters, 2004; Seijts et al., 2004; Chen et al., 2012; Gutman, 2006; Botsas & Padeliadu, 2003; Beck & Schmidt, 2013; Dickhäuser et al., 2021; Janke & Dickhäuser, 2019; Giota & Bergh, 2021).

One challenge to validly interpreting survey responses is that they usually ask learners to aggregate learning experiences across multiple contexts (Turner & Patrick, 2008; Winne, 2010). For example, survey questions often include phrases such as "generally" or "during an exam." This prime is intended to ensure that learners' responses to varying contexts would be consistent for that context. This may not be the case, especially when learners actively self-regulate approaches to learning, as illustrated by research previously cited (see also Hadwin et al., 2001). Moreover, it is usually impractical to administer surveys frequently enough to collect fine-grained data tracking goal changes across the timeline of a single task. When the same or similar questionnaires are given every day, we predict learners acclimate to reporting a generalized "mean experience" rather than taking careful account of varying conditions, particularly if the setting provided in the survey's instructions is not tailored to each administration. While think-aloud or interview methods might lessen this hazard, those methods face other challenges. For example, both for surveys and interviews, learners might respond not based on actual actions but on their knowledge or expectation about which action is recommended for effective learning (Pintrich, 2000). Thus, both accuracy of memory and responses to survey items are in question.

Another issue besetting self-report measures in motivational studies is that learners might not be fully attentive to or willing to report changes in motivational states. Learners' decisions to change goals might be habitual (automated cognition) to the extent that motivation changes go unnoticed. In such instances, fleeting goal changes within a task could be missed in self-reported data. Trace data may be able to complement self-report data in ways that lessen this source of unreliability.

3 Critiques of Recent Studies

In this section, we review three recent studies investigating motivation, each of which collected trace data. We reflect on current methodologies and analyze them to suggest directions for future research. We searched the literature using Google Scholar with three queries: "trace data motivation," "log data motivation," and "clickstream motivation." We then chose reports (1) written in English, (2) using unobtrusive trace data to measure learners' motivation, and (3) published in refereed international conferences or refereed journals. Only a few studies could be identified, all showing commonalities in approaches and making approximately equivalent recommendations for future research. We selected three representative studies for analysis here.

Each study was reviewed using a common schema: theoretical framework, contexts, data and indicators, and data analysis and results. The theoretical framework reveals how tightly the approach in each study connects to motivational theories. Approaches include overall study design, operational definitions of indicators, and interpretations of results in relation to theoretical support. Contexts describe where data were collected. Data and indicators examine types of data collected and which indicators were generated from data. Finally, data analysis and results analyze what and how the findings of each study were drawn.

3.1 Hershkovitz and Nachmias (2008)

The main goal of Hershkovitz and Nachmias' (2008) study was to build a conceptual framework to measure motivation using log data.

3.1.1 Theoretical Framework

From previous literature, the researchers defined three dimensions of motivation: engagement (how strong motivation is), energization (how long motivation is maintained and direction of motivation), and source (if motivation is internal or external). The framework was used to identify indicators relevant to each of these motivational dimensions.

3.1.2 Contexts

Data were collected in a self-paced online system teaching Hebrew vocabulary. Learners could mark each word or phrase as "well known," "not well known," or "unknown" based on familiarity. There were five instructional choices the system offered to learners: memorizing where learners could see words and their meanings, practicing where learners see words without meanings, searching for specific words, gaming, and self-testing.

3.1.3 Data and Indicators

Data analyzed were secondary; that is, the analyses were conducted on pre-existing data not collected specifically for the study. Each row of these secondary log data recorded a session of a learner's activity studying vocabulary. A session was initiated when a learner entered the system and ended when a learner closed the system window. Each log also included attributes such as the start and stop timestamp for each session, the number of words that learners marked as "known," and other actions carried out in the system.

After inspecting raw data for a small number of cases ($N = 5$), the authors identified seven potential indicators of motivation: proportion of time on task, average session duration, pace of actions performed, proportion of words for which learners changed their judgment of familiarity while studying, average time between sessions, proportion of examination events, and proportion of game events.

3.1.4 Data Analysis and Results

Indicators were examined in a larger dataset ($N = 1444$) and reduced by an undescribed method to a final dataset ($N = 674$). Hierarchical clustering was applied, and clusters were mapped based on indicators of the three dimensions of motivation: engagement, energization, and source, defined by the researchers as mentioned above.

3.2 Cocea and Weibelzahl (2011)

These researchers aimed to identify behavioral patterns indicating online learners' motivation levels from log files and ultimately to support low-motivated online learners.

3.2.1 Theoretical Framework

While the researchers reviewed several prior studies investigating particular motivational states such as confidence and effort, their study adopted a general view of motivation as engagement in learning activities. Further distinctions were not a focus in this research.

3.2.2 Contexts

Log data were collected from *HTML -Tutor*, a free online introductory course on HTML. The researchers described the course as interactive but did not provide further details about types of materials or tools, e.g., lecture videos, discussion forums, and the interactive code editor learners used. The amount of material in modules was not described.

3.2.3 Data and Indicators

Timestamped data logged for this research were secondary. Data included events such as login, logout, page access, clicking a hyperlink, using a glossary feature, and searching. From raw log data, the authors created indicators for each participant in the study – performance on tests, time spent reading, number of accessed pages, and time spent on tests – which they used to predict learners' motivation levels. The authors also created a binary indicator of motivation level using rules they established. For example, spending at least 60 s per page on average was considered engaged, while spending less than 20 s per page was categorized as disengaged. This motivation indicator was used to label each learner's overall engagement level as engaged or disengaged. The authors did not report the volume of log data they used to create indicators.

3.2.4 Data Analysis and Results

Log data files of 20 learners were analyzed using the Waikato Environment for Knowledge Analysis (WEKA) system (Witten et al., 1999). In this data analysis, the four indicators except for motivational level were entered into decision trees to predict learners' motivation levels. The motivation level indicator was used as a gold standard to evaluate decision tree predictions. Analysis classified learners as engaged if they spent more than 45 min on reading and showed a performance either above 63% or below 49%.

The authors attributed relatively lower engagement for learners with medium-level performance (between 49% and 63%) to the learners' confidence. The authors interpreted these learners did not invest much effort to improving their performance because the learners judged their level of achievement was good enough.

3.3 *Zhou and Winne (2012)*

The aim of this study was to examine potential differences in achievement goals measured by self-reported surveys and by log data.

3.3.1 Theoretical Framework

Goal orientation theory was the main theoretical framework adopted in this research. The authors criticized self-report measures used in prior research as too divergent in operationalizations of goal orientations. They also questioned whether respondents validly reported goal orientations because self-report items were framed at too coarse a grain size. These researchers designed log data and indicators to capture four goal orientations: mastery-approach, mastery-avoidance, performance-approach, and performance-avoidance. They examined the predictive power of traces of goal orientation as compared to self-report data.

3.3.2 Contexts

Zhou and Winne's study generated primary data in a one-hour-long lab experiment. Learners first responded to the Achievement Goal Questionnaire (Elliot & McGregor, 2001) then read an article about hypnosis. After studying, they took achievement tests posing five multiple-choice items and five short-answer questions.

Two measures of goal orientation were obtained: the self-report Achievement Goal Questionnaire (Elliot & McGregor, 2001) and trace data generated as learners studied in a software system, gStudy. gStudy was a Chrome extension that provided tags and hyperlinks to learners allowing them to choose sources of help to prepare for the achievement test. Each tag and each hyperlink was mapped to one of the four goal orientations according to their labels (e.g., tag: "Reread to avoid misinterpretation" tracing mastery avoidance). Tagging and clicking hyperlinks traced expressions of goal orientations while studying.

3.3.3 Data and Indicators

Zhou and Winne's raw trace dataset was composed of learners' clicks on hyperlinks and tags applied. Counts of traces were used to form four behavioral indicators, one for each facet of goal orientation. For example, if a learner created a tag representing mastery-approach goal orientation five times, that count divided by the total number of all goal orientation traces formed the indicator of mastery-approach goal orientation.

3.3.4 Data Analysis and Results

Results showed correlations between self-reports and trace data were not statistically detectable ($p \geq .05$). Their blocked multiple regression analyses revealed trace-based indicators were statistically better predictors of learners' achievement than any survey-based indicators ($p \leq .01$). Furthermore, all trace-based indicators except one for mastery-avoidance orientation showed a strong Kendall's *tau b* coefficients predicting achievement ($p \leq .01$). None of the survey-based indicators, on the other hand, were a statistically detectable predictor of achievement ($p \geq .05$).

3.4 Critiques of the Select Studies

3.4.1 Importance of Design Processes

Trace data may be noisy, i.e., contaminated with sources of variance not relevant to target constructs. Thus, one important task for researchers is identifying and minimizing noise to enhance the resolution of trace data (Krumm et al., 2022; Winne, 2014). For example, a clickstream datum showing a learner clicked a hint button could indicate various motivation constructs, e.g., simply exploring a software feature vs. attempting to overcome difficulty vs. gaming the system. Noise contaminating trace data, as with any kind of data, jeopardizes valid interpretation. Carefully designing trace data collection in consideration of theories, contexts, and research questions is essential.

In two research cases we reviewed, data were secondary, and the design rationales for motivational indicators were minimally explained. This severely challenges the validity of drawing correspondences between trace data and constructs each trace it intended to represent. Hershkovitz and Nachmias (2008) used secondary data and did not justify how those data represent learners' motivation. They also mentioned they chose indicators used in previous work, but information was minimal about operational definitions as explicit expressions of theory. For instance, their indicator *timeOnTaskPC*, the total time of active sessions divided by the total time logged, was presented as a measure of the engagement dimension of motivation. Because the time learners are logged in can be spent on many different activities, e.g., exploring features of the interface or responding to text messages received on a smartphone, we suggest time on task metrics are typically overly broad and imprecise indicators of motivation.

Similarly, Cocea and Weibelzahl's (2011) use of secondary data prevented designing traces that more directly represent motivational constructs. Furthermore, insufficient explanation regarding their design process limits interpretations of their results. They provided only a table of indicator names and general indicator descriptions. For example, an indicator *NoPages* was described as the number of accessed pages. That indicator is potentially unrepresentative of motivation if a website's architecture requires learners to pass through landing pages or where one website provides a single scrolling page while that same volume of information at another website is distributed across separate pages linked by a "Next" button. Also, it is unclear whether a learner's retreat to a previously viewed page is included in the count *NoPages*. Retreat may be a strong indicator of a learner's motivation to reinstate forgotten information or to monitor clarity about previously studied content.

In contrast, Zhou and Winne (2012) detailed theoretical grounding for designing indicators in their study. While their descriptions might have been more detailed, traces they logged about learners' behavior are explicitly mapped to specific aspects of achievement goal orientation theory according to Elliot and McGregor (2001). This approach permits constructive critique about how those operational definitions manage noise and introduce subjectivity in traces vis à vis constructs they are designed to indicate.

None of these three studies considered motivation change within a learning session even though previous studies show motivation is dynamic (Senko and Harackiewicz 2005). In Hershkovitz and Nachmias's study (2008), the duration of each learner's interaction with the learning platform ranged from 3 weeks to 3 months. Zhou and Winne's (2012) study was just an hour-long and its context was a lab study. Cocea and Weibelzahl (2011) did not clarify how long learners' interaction with *HTML-Tutor* lasted. Methodologies designed to take account of motivational dynamics across the timeline of learner's engagement would be more revealing.

3.4.2 Weak Evaluation Process of Indicators

After generating indicators to measure constructs, it is important to evaluate them in the context of a specific study for future researchers. Two of the three studies we reviewed did not describe an evaluation process: Hershkovitz and Nachmias (2008) did not evaluate their indicators, and Cocea and Weibelzahl (2011) evaluated their indicators by comparing classification results against hand-labeled data identifying whether a learner was engaged or disengaged. While Cocea and Weibelzahl's approach is a step in the right direction, there is no outside criterion beyond the researchers' judgment. As well, some decisions could be considered arbitrary, e.g., choosing "less than 20 seconds spent per page" as the standard for disengagement instead of 15 or 25 s. They chose this threshold based on estimated times for reading a page or working on a test without explaining how these times were estimated. Without sharing further contextual information, such as how many tasks learners had available to work on and some metric of required "steps" to complete each task, it is hard to evaluate the likelihood that indicators of engagement usefully reflect learners' motivation.

Zhou and Winne (2012) evaluated trace-based goal orientation in two ways. First, they examined the correspondence between goal orientations measured by their trace-based indicators and a widely used self-report measure. When they observed weak correspondence between trace and self-report indicators of goal orientation, they examined posttest performance to identify which indicator more strongly aligned to theory's predictions of achievement. They concluded their trace indicators outperformed self-reports as indicators of goal orientation in their study context.

3.4.3 Lack of Discussion on How Trace Measures Were Introduced to Users

While trace data can represent learners' dynamic motivation unobtrusively and, arguably, more directly than self-reported data, benefits may be undermined if the user experience which creates the trace measures is not carefully considered. Traces inherently require the learner to engage with content, e.g., highlight it, or use features in an interface, e.g., a menu of options or a button, controlling software features. If learners are unaware those kinds of engagements are available or do not understand how a software feature functions, trace data will not be generated regardless of learners' motivation, cognition, or metacognition. If the method for creating a trace is perceived to be overly effortful, requires complex maneuvers in the learning environment, or slows the pace of a learner's work too much, learners will avoid the feature that generates trace data. Learners are generally uninterested (and unaware) of the trace data being created, so features of the environment which are instrumented for trace data must provide a clear benefit to the learner in order to be used.

In other words, designing tools to gather trace data requires careful attention to the user experience. In some cases, it may be necessary to provide initial training to learners about how to use trace-generating tools to ensure they understand and appreciate how the tool can be useful in learning. Where the tool appears to learners as a socially desirable property or can be used excessively to game the system, further cautions apply to designing it. We suggest a general guideline: Learning tools which have been designed with tracing methods must have perceived utility to the learners.

Two of the three studies we reviewed did not address the issue of how trace data were related to learner motivational states. For example, in Hershkovitz and Nachmias' (2008) online system teaching Hebrew vocabulary, learners could mark each word or phrase depending on their confidence. Furthermore, learners could use other features such as searching, memorizing, and self-testing. Yet, it is unknown how obvious these features were to learners or how well they were integrated into purposes of the learning task. If data showed learners did not use features after a few attempts or only a few learners continued to use these features, questions arise about the extent to which traces measure enough of behavior and kinds of behavior that serve research goals.

4 Proposals

What does our review of three representative studies suggest for improving online measures of motivation in research and contributing to advancing motivation theories?

4.1 Implementing Design Framework

There are only a few examples of unobtrusive trace indicators of motivation in the field of learning analytics. Researchers aiming to represent motivation using trace data appear likely to design novel indicators rather than build on prior work where strengths and weaknesses of indicators and data designs can be assessed in particular contexts. Thus, we recommend it is important to meticulously inspect each study's design to analyze how and the extent to which it reduces noise and explicitly details key features of the method for generating and logging trace data.

One approach may be using a structured design framework such as the Evidence-Centered Design (ECD) (Mislevy & Steinberg, 2003; Mislevy & Haertel, 2007). ECD is a framework that evaluates assessments designed to permit learners to display knowledge or skills. In this approach, assessment is broadly considered as an argument to be supported by evidence describing learners' latent constructs, such as motivation, knowledge, or a particular skill. Ideally, it would be possible to reliably and validly ascribe a motivational state based on low-noise instances of behavior and patterns.

In particular, ECD's design pattern (Gamma et al., 1995; Alexander et al., 1977) helps researchers build a more solid rationale for their designs of indicators. Implementing a design pattern is often approached by completing a table identifying attributes of a construct and their operational definitions in particular study contexts. For example, suppose a researcher aims to distinguish learners' achievement goals focusing on earning higher final grades (i.e., performance-oriented goals) from mastering learning materials for satisfaction (i.e., mastery-oriented goals). To measure the performance-oriented goal, the researcher designs an indicator as follows: If a learner clicks a hyperlink "critical concepts for the final exam," that could supply evidence of performance-orientation. While implementing the design pattern, the researcher should explain the rationale detailing how this indicator could be strong evidence for performance-oriented goals. Furthermore, the researcher should consider what alternative latent constructs this indicator might represent. For example, learners may click a link simply out of curiosity, not because they hold performance-oriented goals. Through this careful process, a researcher could thoroughly inspect a rationale for a proposed indicator design, potentially improving the design for generating data in ways that improve validity when interpreting data.

Beyond dutiful attention to principles of ECD and considerations Winne (2020a, b) forwarded to improve validity of inferences made and actions (subsequent instructional interventions) based on trace data, we recommend four characteristics for indicators.

First, it should be almost intuitively obvious to learners that information they generate using a tool has value for learning. Highlighted information, for example, eases burdens of locating content judged as meriting review or attention when studying for an examination. Tags greatly facilitate sorting information into categories, e.g., tasks not to be forgotten and major bins in a discipline (e.g., major theorists, disproven hypotheses, useful shortcuts in procedures).

Second, effort required to use a tool should be minimized, thereby reducing extraneous cognitive load. Most undergraduates highlight often and have extensive experience highlighting text in pdf readers or via an extension added to their favorite web browser. Learning how to highlight text once the toolbar icon or keystroke shortcut is introduced is practically one-trial learning. In general, software designs for tools that generate trace data should follow usual guidelines for optimizing the user experience.

Third, the set of tools available to learners should span options for operating on different kinds of information using different operations that achieve different purposes. Without choice, learners are constrained to display variance in their behavior and corresponding inferred underlying processes that comprise SRL. For example, tools for planning steps in a large task and monitoring progress serve quite different purposes than tools for tagging interesting information worth researching further than tools for re-searching information falling into categories.

Fourth, we conjecture learners may be more inclined to "give a tool a chance" if they are provided reasons the tool is designed the way it is. Having and providing a rationale warranting when and why to use a tool may increase chances learners will trial it.

4.2 Evaluating Indicator Designs for Future Studies

To replicate or adopt suggested indicator designs in future studies, it is important to analyze indicators in particular contexts. Construct validity is the degree to which an interpretation of an indicator is justified regarding the presence or degree of a construct. Construct validity is a key concern when evaluating indicators. External validity refers to the degree to which an indicator can be justifiably interpreted in relation to other variables (Messick, 1987). For example, if previous work generally agrees performance goals and academic performance are positively correlated, an indicator designed to capture performance goals should also have a relatively large positive correlation with performance measurements such as posttest scores.

Among studies we reviewed, only Zhou and Winne (2012) correlated trace indicator data purportedly representing achievement goals with posttest scores. That move helps consolidate not only the validity of their indicator designs but also their study's implications. In contrast, Hershkovitz and Nachmias (2008) and Cocea and Weibelzahl (2011) did not pursue these lines of analysis. Combined with a weaker design framework for their indicators, this omission increases uncertainty about the appropriateness of indicator designs in these two studies as a basis for designing future research.

Accumulating evaluation results in diverse contexts is also essential when attempting to generalize motivational indicator designs based on operational definitions of unobtrusive trace data. After particular indicators have been validated as reliably and informatively capturing specific features of learners' motivation in one specific context, those indicators should be examined in related contexts. This would allow researchers to explore for contextual factors affecting the validity attributed to an indicator design.

Researchers should adapt indicator designs to unfolding and varying conditions, both internal and external to the learner. For example, following success on a timed practice quiz problem, learners might be more motivated to choose more difficult problems when they login to the next study session. This motivational change may well affect goals set, tactics chosen, time allocated, and emotional stance. To more accurately capture and analyze such contextual changes implies adapting indicator designs to reflect factors such as changed difficulty levels and new learning tactics. In the abstract, trace data can detect such fine-grained changes but only when researchers forecast changes that may arise and consider how indicators should be re-designed under those changed conditions.

4.3 Introducing Interventions Less Obtrusively

One potential step to reduce noise in laboratory studies is giving learners time to explore and practice using a given system. For example, Zhou and Winne (2012) provided participants with a short practice session, an important opportunity since they asked learners to use unconventional hyperlinks and tags to trace achievement goals. Although brief, the practice session likely increased the chance learners would use these features.

We also suggest researchers consider carefully the context of trace data before including it in an analysis if it was not designed specifically to measure the constructs of interest. While this suggestion does not mean that secondary trace data cannot be used to support analyses, it is important for researchers to consider how that data was created by the system in response to the context of learning. Misaligned data may lead to inappropriate conclusions about leaner motivational states.

Furthermore, we encourage researchers to design features for generating trace data with considerations for learning contexts. Theoretically elegant tools may generate more noise than signal if not tightly articulated to learning objectives and learners' understanding of purposes. For example, a tool learners can use to tag content *research this* generates a clear picture about learners' intentions to engage with additional content. But what is the motive underlying that plan – curiosity, performance orientation (to find material resulting in a higher score on a research paper), anxiety (that important content will be omitted for a research paper)? Steps to usefully constrain interpretations of those trace data, perhaps by changing the label for the tag, may be elusive but necessary.

5 Conclusion

Compared to widely used self-report measures, fine-grained and unobtrusive trace data may often offer stronger alignment to dynamic motivational constructs. Yet, capturing motivational events through trace data remains relatively underexplored in learning analytics, especially how dynamics can be represented to learners and leveraged to guide SRL. Among the few existing studies, rationales for designing indicators of motivation often appear to be insufficiently justified, if at all. This slows advances to theory and curtails the potency of practical recommendations. Our proposals for improving design and validation of indicators that trace constructs should nurture a more rigorous approach to research and the development of more serviceable learning analytics.

References

Alexander, C., Ishikawa, S., & Silverstein, M. (1977). *A pattern language: Towns, buildings, construction*. Oxford University Press.

Ames, C. (1992). Classrooms: Goals, structures, and student motivation. *Journal of Educational Psychology, 84*(3), 261. http://psycnet.apa.org/journals/edu/84/3/261.html?uid=1993-03487-001

Ames, C., & Archer, J. (1988). Achievement goals in the classroom: Students' learning strategies and motivation processes. *Journal of Educational Psychology, 80*(3), 260–267. https://doi.org/10.1037/0022-0663.80.3.260

Beck, J. W., & Schmidt, A. M. (2013). State-level goal orientations as mediators of the relationship between time pressure and performance: A longitudinal study. *The Journal of Applied Psychology, 98*(2), 354–363. https://doi.org/10.1037/a0031145

Botsas, G., & Padeliadu, S. (2003). Goal orientation and reading comprehension strategy use among students with and without reading difficulties. *International Journal of Educational Research, 39*(4), 477–495. https://doi.org/10.1016/j.ijer.2004.06.010

Chen, Z.-H., Liao, C. C. Y., Cheng, H. N. H., Yeh, C. Y. C., & Chan, T.-W. (2012). Influence of game quests on pupils' enjoyment and goal-pursuing in math learning. *Journal of Educational Technology & Society, 15*(2), 317–327. https://www.jstor.org/stable/pdf/jeductechsoci.15.2.317.pdf

Cocea, M., & Weibelzahl, S. (2011). Can log files analysis estimate learners' level of motivation? *Proceedings of the Workshop Week Lernen - Wissensentdeckung - Adaptivitat, Hildesheim*, 32–35.
Dickhäuser, O., Janke, S., Daumiller, M., & Dresel, M. (2021). Motivational school climate and teachers' achievement goal orientations: A hierarchical approach. *The British Journal of Educational Psychology, 91*(1), 391–408. https://doi.org/10.1111/bjep.12370
Elliot, A. J., & McGregor, H. A. (2001). A 2× 2 achievement goal framework. *Journal of Personality and Social Psychology, 80*(3), 501. http://doi.apa.org/journals/psp/80/3/501.html
Elliot, A. J., & Murayama, K. (2008). On the measurement of achievement goals: Critique, illustration, and application. *Journal of Educational Psychology, 100*(3), 613. http://doi.apa.org/journals/edu/100/3/613.html
Elliott, E. S., & Dweck, C. S. (1988). Goals: An approach to motivation and achievement. *Journal of Personality and Social Psychology, 54*(1), 5–12. https://doi.org/10.1037//0022-3514.54.1.5
Fryer, J. W., & Elliot, A. J. (2007). Stability and change in achievement goals. *Journal of Educational Psychology, 99*(4), 700–714. https://doi.org/10.1037/0022-0663.99.4.700
Gamma, E., Helm, R., Johnson, R., Johnson, R. E., & Vlissides, J. (1995). *Design patterns: Elements of reusable object-oriented software*. Pearson Deutschland GmbH. https://play.google.com/store/books/details?id=tmNNfSkfTlcC
Giota, J., & Bergh, D. (2021). Adolescent academic, social and future achievement goal orientations: Implications for achievement by gender and parental education. *Scandinavian Journal of Educational Research, 65*(5), 831–850. https://doi.org/10.1080/00313831.2020.1755360
Gutman, L. M. (2006). How student and parent goal orientations and classroom goal structures influence the math achievement of African Americans during the high school transition. *Contemporary Educational Psychology, 31*(1), 44–63. https://doi.org/10.1016/j.cedpsych.2005.01.004
Hadwin, A. F., Winne, P. H., Stockley, D. B., Nesbit, J. C., & Woszczyna, C. (2001). Context moderates students' self-reports about how they study. *Journal of Educational Psychology, 93*(3), 477–487. https://doi.org/10.1037/0022-0663.93.3.477
Hershkovitz, A., & Nachmias, R. (2008). *Developing a log-based motivation measuring tool* (pp. 226–233). EDM. http://citeseerx.ist.psu.edu/viewdoc/download?doi=10.1.1.464.4775&rep=rep1&type=pdf
Jang, L. Y., & Liu, W. C. (2012). 2 × 2 Achievement goals and achievement emotions: A cluster analysis of students' motivation. *European Journal of Psychology of Education, 27*(1), 59–76. https://doi.org/10.1007/s10212-011-0066-5
Janke, S., & Dickhäuser, O. (2019). A neglected tenet of achievement goal theory: Associations between life aspirations and achievement goal orientations. *Personality and Individual Differences, 142*, 90–99. https://doi.org/10.1016/j.paid.2019.01.038
Kleinginna, P. R., & Kleinginna, A. M. (1981). A categorized list of emotion definitions, with suggestions for a consensual definition. *Motivation and Emotion, 5*(4), 345–379. https://doi.org/10.1007/BF00992553
Krumm, A. E., Coulson, A., & Neisler, J. (2022). Defining productive struggle in ST math: Implications for developing indicators of learning behaviors and strategies in digital learning. In *LAK22: 12th international learning analytics and knowledge conference*. https://doi.org/10.1145/3506860.3506901
Luo, W., Paris, S. G., Hogan, D., & Luo, Z. (2011). Do performance goals promote learning? A pattern analysis of Singapore students' achievement goals. *Contemporary Educational Psychology, 36*(2), 165–176. https://doi.org/10.1016/j.cedpsych.2011.02.003
Masgoret, A.-M., & Gardner, R. C. (2003). Attitudes, motivation, and second language learning: A meta-analysis of studies conducted by Gardner and associates. *Language Learning, 53*(S1), 167–210. https://doi.org/10.1111/1467-9922.00227
Messick, S. (1987). Validity. *ETS Research Report Series, 1987*(2), i–208. https://doi.org/10.1002/j.2330-8516.1987.tb00244.x
Mislevy, R. J., & Haertel, G. D. (2007). Implications of evidence-centered design for educational testing. *Educational Measurement: Issues and Practice, 25*(4), 6–20. https://doi.org/10.1111/j.1745-3992.2006.00075.x

Mislevy, R. J., & Steinberg, L. S. (2003). Focus article: On the structure of educational assessments. *Research and Perspectives*. https://www.tandfonline.com/doi/abs/10.1207/S15366359MEA0101_02?casa_token=VRG8zQQhzU0AAAAA:r4ZnL1_0WgWOglxNoBevXR4IUC7BxGmdG2JH5gz5hq9Qq_XwrCKfwwhNMHzT1BhKj_s15VNyJeEHbA

Muis, K. R., & Edwards, O. (2009). Examining the stability of achievement goal orientation. *Contemporary Educational Psychology, 34*(4), 265–277. https://doi.org/10.1016/j.cedpsych.2009.06.003

Nicholls, J. G. (1984). Achievement motivation: Conceptions of ability, subjective experience, task choice, and performance. *Psychological Review, 91*(3), 328–346.

Pintrich, P. R. (2000). Chapter 14 – The role of goal orientation in self-regulated learning. In M. Boekaerts, P. R. Pintrich, & M. Zeidner (Eds.), *Handbook of self-regulation* (pp. 451–502). Academic. https://doi.org/10.1016/B978-012109890-2/50043-3

Remedios, R., & Richardson, J. T. E. (2013). Achievement goals in adult learners: Evidence from distance education. *The British Journal of Educational Psychology, 83*(Pt 4), 664–685. https://doi.org/10.1111/bjep.12001

Seijts, G. H., Latham, G. P., Tasa, K., & Latham, B. W. (2004). Goal setting and goal orientation: An integration of two different yet related literatures. *Academy of Management Journal, 47*(2), 227–239. https://doi.org/10.5465/20159574

Senko, C., & Harackiewicz, J. M. (2005). Regulation of achievement goals: The role of competence feedback. *Journal of Educational Psychology, 97*(3), 320–336. https://doi.org/10.1037/0022-0663.97.3.320

Tuominen-Soini, H., Salmela-Aro, K., & Niemivirta, M. (2011). Stability and change in achievement goal orientations: A person-centered approach. *Contemporary Educational Psychology, 36*(2), 82–100. https://doi.org/10.1016/j.cedpsych.2010.08.002

Turner, J. C., & Patrick, H. (2008). How does motivation develop and why does it change? Reframing motivation research. *Educational Psychologist, 43*(3), 119–131. https://doi.org/10.1080/00461520802178441

Winne, P. H. (1997). Experimenting to bootstrap self-regulated learning. *Journal of Educational Psychology, 89*(3), 397. http://psycnet.apa.org/fulltext/1997-05647-001.html

Winne, P. H. (2010). Improving measurements of self-regulated learning. *Educational Psychologist, 45*(4), 267–276. https://doi.org/10.1080/00461520.2010.517150

Winne, P. H. (2014). Issues in researching self-regulated learning as patterns of events. *Metacognition and Learning, 9*(2), 229–237. https://doi.org/10.1007/s11409-014-9113-3

Winne, P. H. (2020a). Commentary: A proposed remedy for grievances about self-report methodologies. *Frontline Learning Research*. https://eric.ed.gov/?id=EJ1260776

Winne, P. H. (2020b). Construct and consequential validity for learning analytics based on trace data. *Computers in Human Behavior, 112*, 106457. https://doi.org/10.1016/j.chb.2020.106457

Winne, P. H. (2022). Modeling self-regulated learning as learners doing learning science: How trace data and learning analytics help develop skills for self-regulated learning. *Metacognition and Learning*. https://doi.org/10.1007/s11409-022-09305-y

Winne, P. H., & Marzouk, Z. (2019). Learning strategies and self-regulated learning. In J. Dunlosky (Ed.), *The Cambridge handbook of cognition and education* (Vol. 729, pp. 696–715). Cambridge University Press, xviii. https://doi.org/10.1017/9781108235631.028

Witten, I. H., Frank, E., Trigg, L. E., Hall, M. A., Holmes, G., & Cunningham, S. J. (1999). *Weka: Practical machine learning tools and techniques with Java implementations*. https://researchcommons.waikato.ac.nz/handle/10289/1040

Wolters, C. A. (2004). Advancing achievement goal theory: Using goal structures and goal orientations to predict students' motivation, cognition, and achievement. *Journal of Educational Psychology, 96*(2), 236–250. https://doi.org/10.1037/0022-0663.96.2.236

Zhou, M., & Winne, P. H. (2012). Modeling academic achievement by self-reported versus traced goal orientation. *Learning and Instruction, 22*(6), 413–419. https://doi.org/10.1016/j.learninstruc.2012.03.004

Chapter 6
Measuring Collaboration Quality Through Audio Data and Learning Analytics

Sambit Praharaj, Maren Scheffel, Marcus Specht, and Hendrik Drachsler

Abstract Collaboration is an important twenty-first-century skill. Collaboration quality detection can help to support collaboration. This chapter addresses the collaboration quality detection and measurement: (1) to define collaboration quality using audio data and unobtrusive learning analytics measures; (2) to explain the design of a sensor-based set up for automatic collaboration analytics; (3) to move toward quantifying the quality of collaboration by using this set up and show the analysis using meaningful visualizations. Furthermore, we address the challenges and issues at hand and how solutions can be built upon the work already done. To elaborate the different chapter's objectives, we use the terminology of indicators (i.e., the events) and indexes (i.e., the process) to define the components to detect collaboration quality. In one study, during collaborative brainstorming, higher was the equality (i.e., the index) of total speaking time (i.e., the indicator), lower was the dominance of each group member (in terms of total speaking time), and better was the quality of collaboration. However, quality of collaboration is dependent on the

S. Praharaj (✉)
Center for Advanced Internet Studies, Bochum, NRW, Germany

Ruhr-Universität Bochum, Bochum, NRW, Germany
e-mail: sambit.praharaj@cais-research.de

M. Scheffel
Ruhr-Universität Bochum, Bochum, NRW, Germany
e-mail: maren.scheffel@rub.de

M. Specht
Delft University of Technology, Delft, South Holland, The Netherlands
e-mail: m.m.specht@tudelft.nl

H. Drachsler
DIPF Leibniz Institute for Research and Information in Education, Frankfurt am Main, Hessen, Germany

Goethe-Universität, Frankfurt am Main, Hessen, Germany

Open University of the Netherlands, Heerlen, Limburg, The Netherlands
e-mail: h.drachsler@dipf.de

V. Kovanovic et al. (eds.), *Unobtrusive Observations of Learning in Digital Environments*, Advances in Analytics for Learning and Teaching,
https://doi.org/10.1007/978-3-031-30992-2_6

context of collaboration and the actual content of the discussion. During collaboration content analysis has been mostly on the surface level by using certain representative keywords to model different topic clusters. Therefore, we develop a sensor-based setup for automatic collaboration analytics to understand collaboration quality holistically in a learning context. Here, our aim is to understand "how" group members speak (i.e., speaking time indicator) and "what'" (i.e., the content of the conversations) group members speak to move toward collaboration quality measurement.

Keywords Collaboration analytics · Collaboration quality · Learning analytics · Group work · Technology-enhanced learning · Multimodal learning analytics

1 Introduction

Collaboration is an important twenty-first-century skill (Dede, 2010) and one of the 4Cs skill set along with critical thinking, communication, and creativity (Kivunja, 2015). Collaboration is said to occur when two or more people work toward a common goal (Dillenbourg, 1999). Most of the works in the field of learning analytics about support for collaboration have focused on analyzing remote (or online) collaboration (Jeong & Hmelo-Silver, 2010). However, with the widespread adoption of sensors (Grover et al., 2016; Kim et al., 2008), multimodal learning analytics (MMLA) (Blikstein, 2013; Di Mitri et al., 2018; Praharaj et al., 2018a) has gained prominence, thus redirecting attention to the analysis of co-located collaboration (CC) (or face-to-face collaboration) with the help of sensor technology (Grover et al., 2016; Kim et al., 2008; Praharaj et al., 2021b; Tausch et al., 2014). Moreover, sensor technology can be easily scaled up (Reilly et al., 2018) and has become affordable and reliable in the past decade (Starr et al., 2018). CC takes place in physical spaces where all group members share each other's social and epistemic space (Praharaj, 2019). Social space is composed of the non-verbal interactions (such as change in posture and specific gesture) and the non-verbal audio interactions (such as total speaking time and turn-taking). Epistemic space comprises the verbal audio interactions (such as the actual content of the conversations).

Collaboration is a complex process. "The requirement of successful collaboration is *complex, multimodal, subtle*, and learned over a lifetime. It involves *discourse, gesture, gaze, cognition, social skills, tacit practices*, etc." (Stahl et al., 2013, pp. 1–2, emphasis added). Meier et al. (2007) identified five facets of collaborative process and nine dimensions of rating collaboration quality: communication (sustaining mutual understanding, dialogue management), joint information processing (information pooling, reaching consensus), coordination (task division, time management, technical coordination), interpersonal relationship (reciprocal interaction), motivation (individual task orientation). A collaboration activity can be called successful or not depending on the focus of the assessment of collaboration, i.e., whether collaboration is assessed as a process or as an outcome (Child & Shaw, 2015).

To measure how successful a collaborative activity is, we need to detect the quality of collaboration. Quality of CC can be detected by different indicators (i.e., the events) of collaboration such as total speaking time (Bachour et al., 2010) or eye gaze (Schneider et al., 2015). These indicators after processing and aggregation can be grouped into different indexes (i.e., the process) which act as the measurable markers of CC quality. For example, the quality of collaboration within a group can be good if there is higher equality (i.e., the index) of total speaking time (i.e., the indicator) among the group members (Bachour et al., 2010). Furthermore, different scenarios of CC such as collaborative programming (Grover et al., 2016), collaborative meetings (Kim et al., 2008; Terken & Sturm, 2010), or collaborative brainstorming (Tausch et al., 2014) each has a different set of indicators denoting the quality of collaboration. For instance, in collaborative programming relevant indicators of collaboration include pointing to the screen, grabbing the mouse from the partner, and synchrony in body posture (Grover et al., 2016); whereas in collaborative meetings gaze direction, body posture, or speaking time of group members are more relevant indicators for collaboration quality (Kim et al., 2008; Stiefelhagen & Zhu, 2002; Terken & Sturm, 2010). This difference can be attributed to the goals of the collaborative tasks and the group characteristics.

While defining indicators and indices represents the first step in measuring the quality of face-to-face collaboration, another significant challenge is the automated capturing of indicators in a scalable manner. In our work, we focus mainly on audio data, because it was the most used modality in the past studies. It can be attributed to the ease of capturing audio with a very minimalistic setup like a microphone. The CC quality has been detected from simple audio indicators of collaboration such as total speaking time and indexes like equality of total speaking time (Bachour et al., 2010; Bergstrom & Karahalios, 2007). Focus of most studies in the past was on "how group members talk" (i.e., spectral, temporal features of audio like pitch) and not "what they talk". The "what" of the conversations is more open, contrary to the "how" of the conversations in understanding what happened during collaboration (Praharaj et al., 2021b). Very few studies studied "what" group members talk about, and these studies were lab-based showing a representative overview of specific words as topic clusters (Chandrasegaran et al., 2019) instead of analyzing the richness of the content of the conversations by understanding the linkage between these words.

To overcome this, we made a starting step based on field trials to prototype, design a technical set up to collect, process, and visualize audio data automatically. The data collection took place while a board game was played among the university staff with pre-assigned roles to create awareness of the connection between learning analytics and learning design. We not only did a word-level analysis of the conversations, but also analyzed the richness of these conversations by visualizing the strength of the linkage between these words and phrases interactively. In this visualization, we used a network graph (Praharaj et al., 2021b) to visualize turn-taking exchange between different roles along with the word-level and phrase-level analysis. This helped us to move toward automated collaboration quality detection.

Therefore, the focus of the chapter is to provide an overview of unobtrusive measures of collaboration quality (in Sect. 2) with the help of a literature review where we define the collaboration quality. Then we provide an outline of one particular method that is based on audio data. Thus, in Sect. 3, we explain the weakness of the past studies using audio data. In Sects. 4, 5, and 6, we explain our approach to move toward automated collaboration quality detection by using analytics, visualizations, and then to finally give meaningful feedback. In Sect. 7, we discuss the challenges and then in Sect. 8, we have a broader discussion, conclusion, and recommendations for future researchers in the field.

2 Defining Collaboration Quality

Collaboration quality helps us to ascertain whether a collaborative activity was successful or not. Collaboration quality is defined based on our literature review (Praharaj et al., 2021a). The broader objective of the review was to find the co-located collaboration (CC) indicators that have been detected using different modalities (such as audio, video) to understand the quality of CC.

In the first round of the analysis during the literature review, the selected publications were classified according to the sensors, indicators, and indicator types as in Fig. 6.1. One or more indicator types can be tracked using the data streams from the sensors and processing them. For instance, a microphone sensor can only track

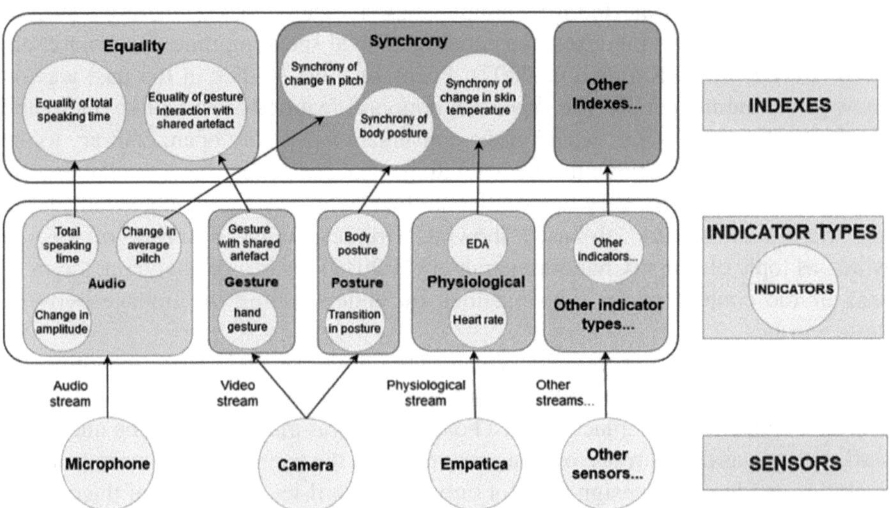

Fig. 6.1 Outline for the terminology used in the review (i.e., sensors, indicators, indicator types, and indexes) to define collaboration quality. (Reprinted from Praharaj et al. 2021a)

audio indicator type using the audio data stream whereas multiple indicator types like audio, posture, gesture, and spatial can be tracked by a Kinect (i.e., an integrated sensor which can simultaneously act as an infrared, depth, audio and video sensor). Each indicator-type cluster is composed of multiple indicators of CC detected by the sensors. For example, audio data is composed of different indicators such as pitch, amplitude, and speaking time detected by the microphone sensor.

The indicators when processed and aggregated can then be grouped to high-level indexes which define the quality of collaboration. For instance, a group which shows higher *equality* (i.e., the index) of total speaking time (i.e., the indicator) during CC has a better quality of collaboration (Bachour et al., 2010; Bergstrom & Karahalios, 2007). In the literature review, we discuss the different indicators, indicator types, and indexes of collaboration quality in-depth in more than 80 different studies with different tables which is not in the scope of this chapter. Here, we limit ourselves to the conceptual definition of collaboration quality.

But, speaking time cannot be a good indicator of collaboration across all the different scenarios of collaboration (such as collaborative programming, collaborative brainstorming, collaborative problem solving). For different scenarios, indicators of collaboration quality vary (Praharaj et al., 2018b) depending on the context. Thus, we made a scenario-driven prioritization to choose a set of indicators depending on the particular scenario of CC in the review. This formed the basis for modeling the collaboration detection framework by mapping the fundamental parameters in those scenarios onto the indicator types and indexes. There are different fundamental parameters in each scenario because of differing goals of different scenarios, team composition (such as roles and compulsory interaction with specific artifacts because of the task type), and varied group behavior (such as dominance or coupling). For example, some CC tasks already have pre-assigned roles (Hare, 1994) for each group member and in some tasks, roles emerge during collaboration (Strijbos & Weinberger, 2010). Some group members are more dominant while others are not.

Figure 6.2 shows the main outcome of the review as to how collaboration quality is detected using both the bottom-up approach (starting from the data streams of the sensors) and then the top-down approach (starting from the different scenarios of collaboration).

The mapping of these goals and parameters to the indicators and indexes to detect collaboration quality has been discussed in-depth in the literature review (Praharaj et al., 2021a) with tables. For the scope of this chapter, we will give one example from it. For example, if there is less dominance (i.e., the parameter) in the group then synchrony (i.e., the index) in body posture (i.e., the indicator) is high and the quality of collaboration is good which basically means that not one member is actively changing the posture to do the task, but everyone is actively or passively contributing to it (Kim et al., 2008).

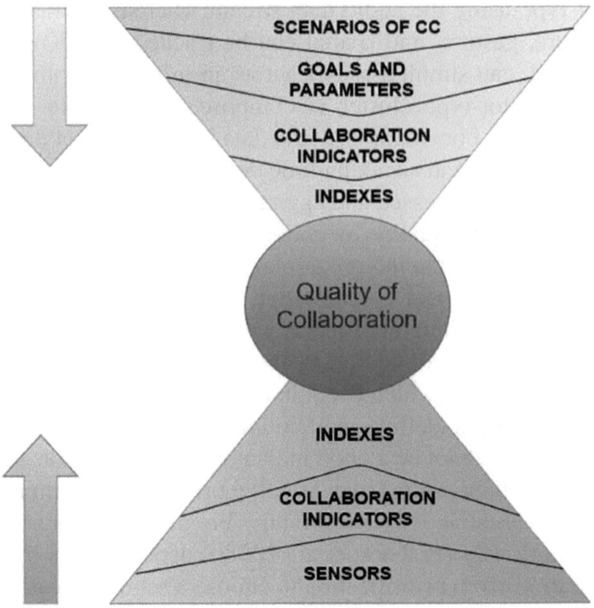

Fig. 6.2 CC quality detection using both bottom-up and top-down approach. (Reprinted from Praharaj et al. 2021a)

3 Background

We narrow our focus on group audio indicator type to detect collaboration quality only because of the abundant availability and ease of audio data collection (Praharaj et al., 2021a). Apart from the majority of studies focusing on the analysis of *how* group members speak (for instance, speaker-based indicators like the intensity, pitch, and jitter were used to detect collaboration quality among working pairs (Lubold & Pon-Barry, 2014)), very few studies used the *what* (or the content) of the audio for the analysis of CC quality.

For example, the "talk traces" (Chandrasegaran et al., 2019) and "meeter" (Huber et al., 2019) studies analyzed the content of the conversation. In the "talk traces" study, Chandrasegaran et al. (2019) did topic modeling during the meeting and then showed the topic clusters as visualization feedback by comparing with the meeting agenda. Furthermore, topic modeling is based on a collection of representative keywords which barely scratches the surface. It does not show the proper connection between these words and the rest of the conversation, which can lead to the loss of the holistic meaning of the conversations and a possible under-representation of certain topics.

The "meeter" study (Huber et al., 2019) classified the dialogues of the group members based on a lab study to measure information sharing and shared understanding while generating ideas. The collaborative task was based on three open-ended fixed topics where group members needed to brainstorm and share their ideas in a short session of 10 min. Their performance (or the quality of collaboration) was measured based on the number of ideas they wrote down on the cards, which was quality controlled before counting the total ideas to remove bad ideas. They did not find significant effects of information sharing and shared understanding on the quality of collaboration. Therefore, the studies analyzing the content of the conversations were too abstract and mostly lab-based. To overcome these limitations, we conducted field trials to build a technical setup and then prototyped it in real-world settings to move toward automated collaboration analytics from group speech data.

Table 6.1 shows an overview of the indicators of CC and their operationalization using the group audio data in some past studies. CC takes place in physical spaces

Table 6.1 Indicators of CC and their operationalization of collaboration quality

Parameters	Indicators	Operationalizing collaboration quality	Space tracked	References
Roles (leader and follower)	Topics covered (topics are detected from keyword clusters and phrases)	Topical closeness to meeting agenda, role-based usage of keywords	Epistemic	Chandrasegaran et al. (2019) and Praharaj et al. (2021b)
Dominance	Total speaking time	Higher equality of total speaking time means less dominance and higher quality of collaboration	Social	Kim et al. (2008), Bachour et al. (2010), Bergstrom and Karahalios (2007), Praharaj et al. (2019)
Active participation	Turn-taking frequency	Frequent turn-taking changes mean higher active participation and better quality of collaboration	Social	Kim et al. (2015)
Expertise	Overlapped speech	Overlap in speech is an indicator of constructive problem solving, expertise, and good CC quality	Social	Zhou et al. (2014) and Oviatt et al. (2015)
Rapport	Synchrony in rise and fall of average pitch	Higher synchrony in rise or fall of average pitch indicates higher rapport and CC quality	Social	Lubold and Pon-Barry (2014)
Knowledge co-construction	Knowledge convergence (i.e., the amount of shared knowledge in the group), cognitive convergence	Increase in convergence (i.e., increase in shared knowledge) implies increase in CC quality	Epistemic	Jeong and Chi (2007) and Teasley et al. (2008)

Adapted from Praharaj (2022)

at the intersection of the group members' social and epistemic space (Praharaj, 2019). The *social* space consists of *how* group members speak, and the *epistemic* space consists of *what* they speak.

4 Automated Collaboration Analytics

To overcome the challenges, we did a field study where we looked at both the spaces to get a holistic overview of the collaboration analytics. We used the Fellowship of Learning Activity (FOLA^2) (http://www.fola2.com/, last accessed on 17 April 2023) board game where university staff with pre-assigned roles (such as teachers, all advisors (consisting of learning analytics advisor and educational advisor), learners, study coach, and game master) designed a learning activity. The main objective of this game is to create awareness of the connection between learning analytics and learning design. This game was played with different themed cards to steer the discussion in different phases for around 60–90 min in each session. In each phase, the cards had keywords related to that phase which were shown by the game master one after the other as the discussion progressed. For example, in technology phase-related discussions there were cards on interaction technologies like shakespeak and powerpoint. There were a total of 14 sessions where we recorded the audio data during the collaborative game design sessions and all these discussions were in the Dutch language. For this recording, we used clip-on microphones attached to each group member which recorded audio to the local recorder attached to those microphones.

After each game design session, these audio files were immediately transferred to the central storage space, which was the long-term storage. For the pre-processing and subsequent operations on the data, we took a copy of the files in the storage space for the pre-processing and processing unit. Here, we pre-processed and transcribed these audio files using Amber Script (https://www.amberscript.com/en/, last accessed on 28 Nov 2022). Finally, the data were processed using Natural Language Processing and analyzed to generate meaningful insights and passed on to the visualization unit to generate the visualizations. These visualizations were generated in a post hoc manner after the group meetings.

The data pre-processing, processing, analysis, and visualizations were done in Python using different openly available libraries. We pre-processed the stored audio files for each group member by extracting the timestamps from the audio file (in .wav audio file format), did speaker diarization (i.e., "who spoke when?"), and then transcribed it at the same time. Finally, we made a .csv file which contains the transcribed text, timestamps, and the roles of who spoke that text at which time. Figure 6.3 shows the data table in CSV file format after pre-processing.

This table was used to analyze the content of the conversation across sessions and role-to-role exchanges with time. We used *natural language processing* in Python for analyzing the text which includes cleaning, processing, and analyzing the text. This helped us to build the text corpus for analysis and visualizations. The following steps helped in cleaning the data:

	Start time in millisecond	End time in milliseconds	Names	Text_y
0	0	1759		
1	1760	1919	256	
2	1919	2079	256	
3	2080	23599	Game Master	Nou de voorbereidingen gedaan dan Ja zijn allemaal willekeurig Ja ik heb ook niet zoveel bij iemand iets neergelegd ik heb even gezorgd dat de student en docent dat die al
4	23600	23679	256	
5	23680	29759	Game Master	hebben ze dat makkelijk voor jullie liggen ze daar ook ja gewoon
6	29760	29839	256	
7	29840	30639	Game Master	
8	30640	30719	256	
9	30720	37279	Game Master	Google dahua wel een hoesje omheen zit mag ook echt opschrijven en stiften
10	37280	37359	256	
11	37360	54319	Game Master	watervaste en whiteboard Marker stift okÃ© stop doe ik eerst even met je moet ik eerst even met jullie over praten hebben we het vandaag over discussie College van we
12	54320	55599	TELI LA Advisor	onderuit
13	55600	56879	Educational Advisor	Varsseveld
14	56880	59039	Game Master	misten
15	59040	61599	TELI LA Advisor	ja ja
16	61600	61679	Game Master	
17	61680	61759	256	
18	61760	70079	Game Master	individueel activiteiten
19	70080	74479	Teacher	en wat
20	74480	76319	Game Master	staat ook achter op het bord
21	76320	79759	TELI LA Advisor	de vier onderwerpen die in dat
22	79760	112559	Game Master	College langskomen zijn de onderwerpen 10 Focus teamrollen zijn is het onderwerp je eigen piemel verder ervaring in het kader van de belbin rollen en beide rollen beide is
23	112560	112639	256	
24	112640	134479	Game Master	beschrijving Bel Mauritsschool zoveel op dat onderdeel voor het onderdeel zo voordat onderdelen onderdelen dus dat is eigenlijk het tweede onderwerp ervaring wordt no
25	134480	134559	256	
26	134560	135199	Game Master	

Fig. 6.3 The stored data table sample

- Tokenization—The process of splitting the sentences into good words or tokens. It lays the foundation for the next steps of cleansing.
- Elimination of stop words—The process of removing words that mean little; these are usually words that occur very frequently. Apart from using the libraries in Python for stop word removal, we also defined our list of contextual stop words libraries that were considered unimportant for this model.
- Lemmatization and stemming—Lemmatization and stemming convert a word into its root form. For example, for the words "running" and "runs", the stem of both words is run. Thus, after we stemmed, these words would be grouped together and retain the same meaning for the model even though they had different forms.
- Sentence segmentation—We split the unstructured spoken text into different sentences, which helped the model understand the boundaries of the long text to make it more semantically distinct.
- Vectorization—Since we cannot input plain words into a model and expect it to learn from it, we had to vectorize the words. We encoded words using high-dimensional vectors where the different dimensions of vectors represent the latent meaning of the words. Therefore, the vectorized version of words would be useful later while generating bigrams (two-word combinations appearing together), trigrams (three-word combinations appearing together), and topic modeling based on the keywords or grouping semantically similar keywords.

The processed data can be used to generate different analytics and visualizations to get insights about the collaboration processes during collaborative game design.

5 Toward Collaboration Quality Detection: From Analytics to Visualizations

First, we do an exploratory analysis and visualization on the processed text data. We use topic modeling with Latent Dirichlet Allocation and Latent Semantic Indexing and then visualize the representative keywords showing different topics in one phase of one session where the main discussion is supposed to be about technology. Figures 6.4, 6.5, and 6.6 show an overview of the topics.

Topic 1 dealt with the use of different types of interaction technology as discussed in this phase. These were mainly evident from the words: “technologie”, “shakespeak”, “sendstep”, and “smart”. These technologies were to be used by the teacher while interacting with the learner, which was evident from the word “docent”, which means “teacher” in English. On examining further, the advisors (supposed to discuss technology and learning analytics) had a higher probabilistic likelihood of getting topic 1. Topic 2 refers to the use of moodle for assignments, making a photo of the post-its using the phone. This topic cluster also captured bad (“slecht”) teams, ideas, and overview roles (“rol”) per student. The last topical

Fig. 6.4 Topic 1: Interaction technologies

Fig. 6.5 Topic 2: Using moodle for assignments

Fig. 6.6 Topic 3: Using red cards on technology

cluster, Topic 3, focused on the use of red cards ("rod", "kaart") (or cards supposed to be used to discuss technology) and learning technology ("leertechnologie"). Then we observe the role-based bigrams and trigrams to find the interesting discussions temporally in each session. The details of the bigrams and trigrams discovered can be found in Praharaj et al. (2021b).

To do an in-depth holistic analysis of collaboration quality, we analyze both the social and epistemic space. First, we visualize the *total speaking time* and *turn-taking* from the social space and then we visualize the *content of the conversations* from the epistemic space as in Praharaj et al. (2021b). For visualizing the social space, we take the help of a node-edge network graph where each node shows a group member with a certain role and the edge shows the turn-taking between the members as in Figs. 6.7 and 6.8. The size of the node is proportional to the total speaking time of that role and the thickness of the edges is proportional to the number of turn-taking exchanges between the roles. This can help us to understand the dominant role-role exchanges temporally so that we know how the conversation patterns evolve with time.

Then, it will be interesting to visualize the epistemic space as to why certain roles have more turn-taking and dominate the conversation. Is it collaborative task-related discussion or is it clarification about the role-based tasks? To understand this further we first visualize the epistemic space to show the role-based usage of frequently used keywords during collaboration temporally. Figures 6.9 and 6.10 show the role-based usage of frequently uttered keywords in the first 20 and 30 min of the first session respectively. This helps us to understand how the usage of specific content-related or unrelated keywords is used by different roles and how it changes with time.

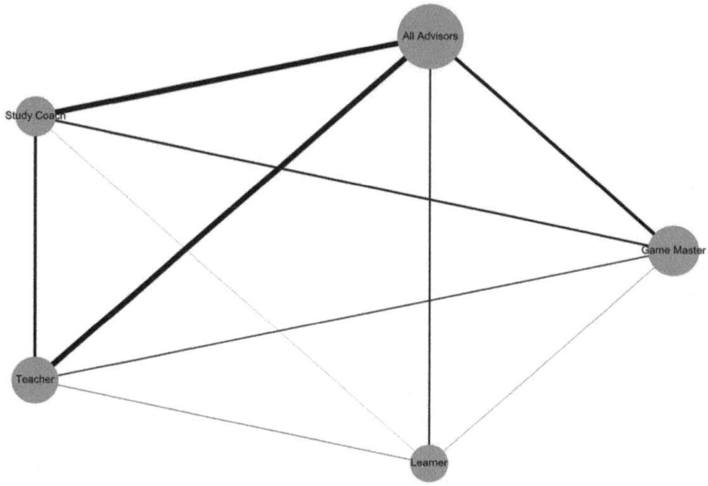

Fig. 6.7 First 20 min of social space in the first session. (Adapted from Praharaj et al. 2022)

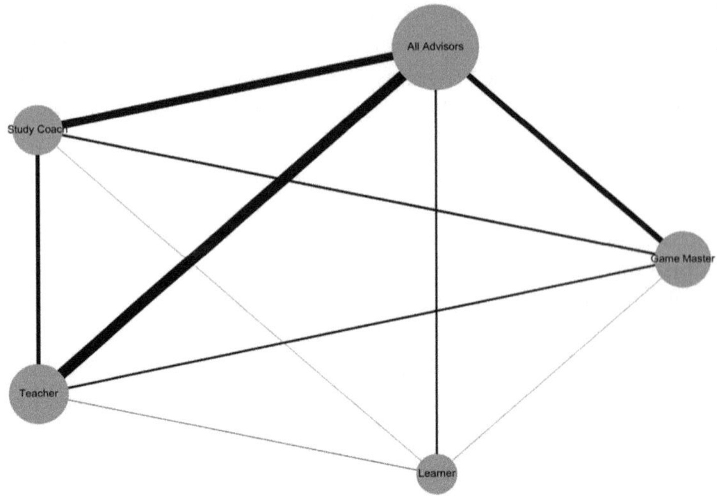

Fig. 6.8 First 30 min of social space in the first session. (Adapted from Praharaj et al. 2022)

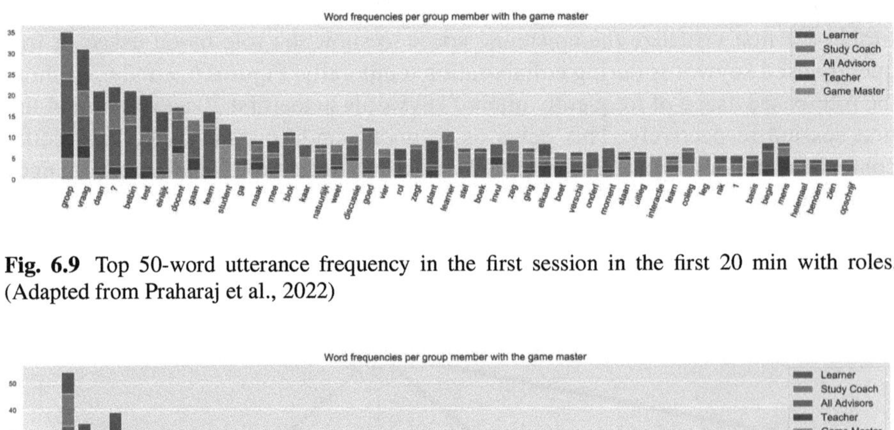

Fig. 6.9 Top 50-word utterance frequency in the first session in the first 20 min with roles. (Adapted from Praharaj et al., 2022)

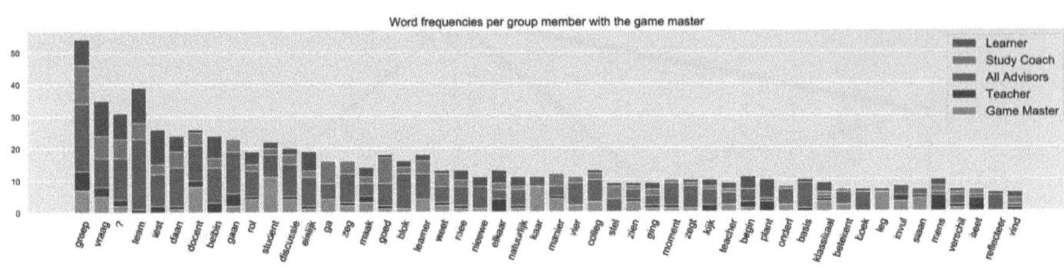

Fig. 6.10 Top 50-word utterance frequency in the first session in the first 30 min with roles. (Adapted from Praharaj et al., 2022)

Furthermore, we used the concept of knowledge convergence to quantify the quality of collaboration, i.e., how the shared knowledge among the group members (with different roles) changes as measured by the usage of different keywords with time. For instance, in Fig. 6.11, "team", a context-relevant keyword isn't spoken by the teacher in the first 10 min of the conversation but then in the next 10 min, i.e., in

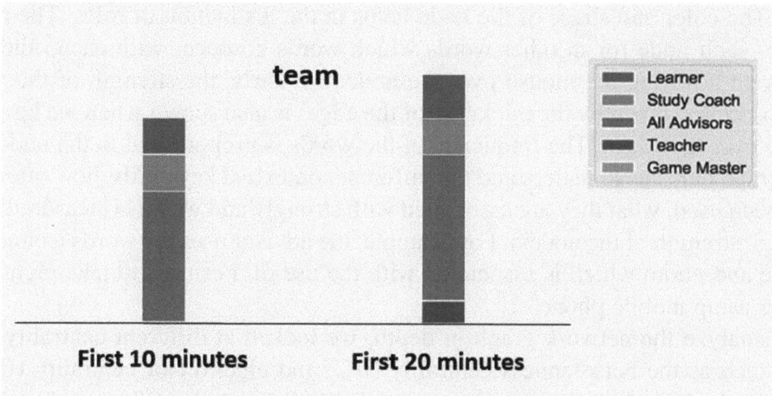

Fig. 6.11 Knowledge convergence example

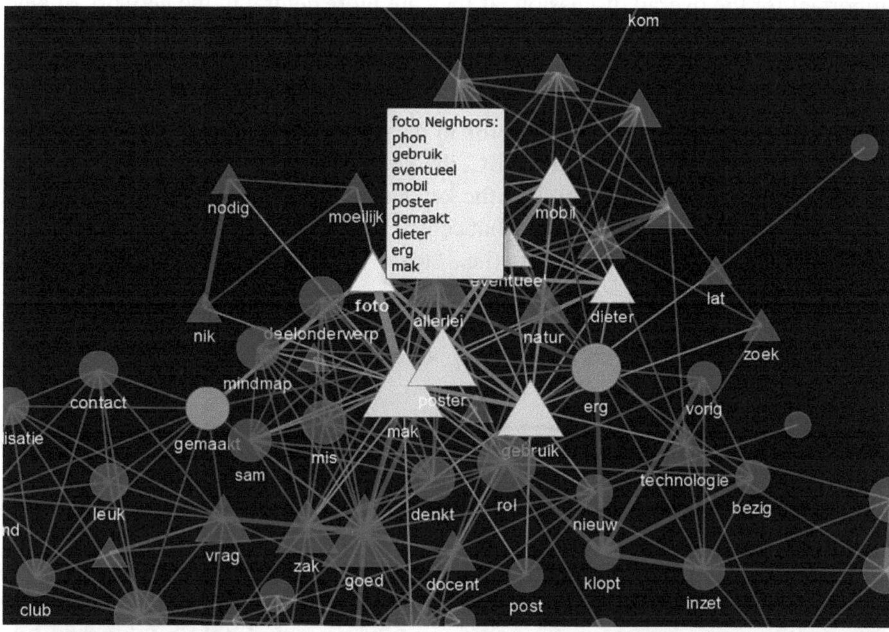

Fig. 6.12 Zoomed-in network graph highlighting a node of the advisor in rectangles and rest others in circles in technology phase of a session. (Adapted from Praharaj et al. 2022)

the first 20 min, the teacher also becomes part of the shared knowledge space of the team keyword. This signals an increase in shared context-relevant keyword knowledge convergence and thereby an increase in the quality of collaboration.

Moving from keywords to the phrases, we visualized how different words co-occur in a sentence using the network graph as in Fig. 6.12. This figure shows a zoomed-in version of the advisor role among other roles with different shape and color. The color and shape of the node helps in the distinction of roles. The neighbors of each node (or in other words which words co-occur with each other) are shown on hovering the mouse over the node. Similarly, the strength of the words that co-occur (shown by the thickness of the edge) is also shown when we hover the mouse over the edges. The frequency of the words is proportional to the node size. This graph helps us to understand the different contextual keywords, how often they have been used, what they are associated with strongly and weakly (measured by on the edge strength of the nodes). For example, the advisor uses the words technology, mobile and photo which is associated with the use of a camera to take pictures of posters using mobile phone.

To analyze the network graph in depth, we looked at different centrality measures such as the betweenness centrality (BC) and eigenvector centrality (EC) of these words. Betweenness centrality shows how often a node (or keyword) acts as a bridge node, that is the number of times a node lies on the shortest path between other nodes. This means that keywords with high betweenness centrality are more important for the overall discussion, as they are more central in the network of keywords. Eigenvector centrality indicates the influence of a node. Therefore, a node with a high eigenvector centrality score must be connected to many other nodes who themselves have high scores. For example, in the technology discussion phase of the first session, frequency wise four words in decreasing order were "good", "make", "moodle", and "use". But, based on BC, the key terms were "good", "team", "use", and "technology", and based on EC, the key terms were "make", "poster", "good", and "role". So, this example shows that centrality measures can elevate the ranking of even less frequently used words (i.e., "team", "technology", and "role" in this example) in that context.

Figure 6.13 provides a holistic overview of the collaboration from group audio data. It shows the dashboard highlighting a node for all advisors in the technology

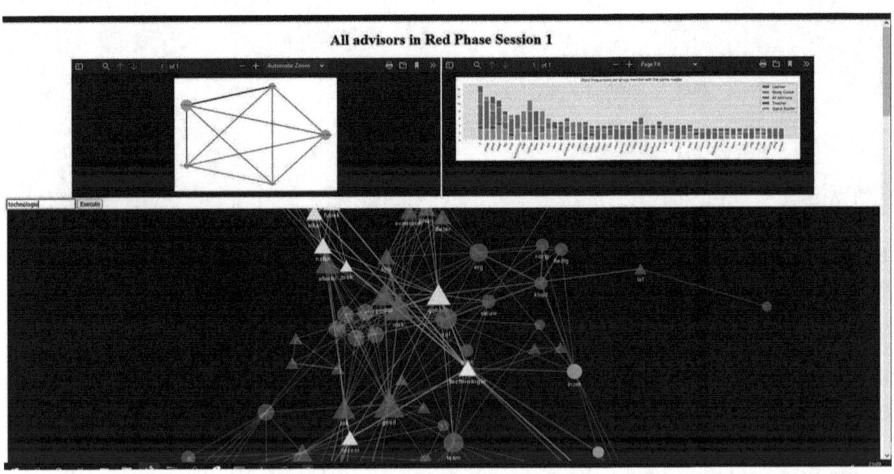

Fig. 6.13 Screenshot of the dashboard with social and epistemic components. (Adapted from Praharaj et al. 2022)

(or red) discussion phase in session 1. It has four main parts. The social space is shown by the role network graph. The high-level overview of the epistemic space is shown by the bar graph which shows role-based usage of the keywords. The colorful network graph shows the interaction of a particular role in one phase of a session. Finally, the search bar which helps to search and highlight a specific node (which is also possible on clicking on that node). Now we have different views for each phase and session with each view showing the conversation of one role in the whole conversation network graph. This will make it easier to compare two roles' conversation patterns when they are seen side by side. This dashboard is scalable, dynamic, and interactive.

6 From Visualizations to Meaningful Feedback

We will build a generic dashboard (taking help of the dashboard prototype) to quantify collaboration quality based on different collaboration indicators in the social and epistemic space with different visualizations. This dashboard will be useful to show how each role interacted during the collaboration task temporally, who was dominating the task. Now, the important question is: "Who would use it and why?". This question will be answered by understanding the needs of the dashboard design.

The design of the dashboard will be driven by the temporal needs (i.e., whether updated in real-time every few minutes or shown as a summary at the end of collaboration) and the stakeholders (teacher or task moderator or the group members themselves) who will be using it.

To address the temporal needs, we need to first differentiate what can be shown as immediate formative feedback and what can be shown as summative feedback at the end of collaboration. To this end, we need to do a qualitative study by interviewing different stakeholders to identify the user requirements. This will give us an idea as to what type of feedback is relevant for which stakeholder group and can be shown to them accordingly. For instance, this type of dashboard for a teacher (as the stakeholder) could be useful to determine scaffolding strategies during collaboration and also planning the collaboration sessions. The pedagogical meaning should be clear for the teacher to act as meaningful feedback. Is it relevant to show continuously who is dominating based on the speaking time and turn-taking or is it relevant to show certain triggers for the teacher to act like suppose when group members are confused or spending too much time in off-topic discussions? For the group members (as a stakeholder), it can be a useful tool to self-reflect (when the feedback is like a mirror) and adapt their collaboration accordingly. It might also be a more advanced version of AI-driven feedback which prompts the group members to act or behave in a certain way to enhance their collaboration.

These are some of the questions that need to be taken care of when customizing the dashboard for different stakeholders. Based on that we can also do design enhancements and modifications in the dashboard using different visualization filters to capture and compare temporal role-based snapshots.

7 Challenges

First, there are theoretical challenges. In some studies, indicators are used directly to understand the quality of collaboration without aggregating them to indexes or understanding how they contribute to collaboration quality. For example, silence has been used as an indicator of collaboration quality without understanding if more or less silence is good for the quality of collaboration. In those examples, silence was used as a feature for machine learning classifiers along with other indicators of collaboration to compute the quality of collaboration. Therefore, operationalization of the indexes to determine CC quality suffers from coding complexity even though many exist on a theoretical level (such as mutual understanding, information pooling, and others as in Meier et al. (2007)). So, there needs to be more adoption of these indexes to bring them into practice to test their strengths and limitations to understand the quality of collaboration.

Next, technical challenges are the degree of automation and the accuracy of speech to text transcription. There are challenges in processing and analyzing the data, which are largely dependent on the input (i.e., the transcribed data). The unstructured text data obtained from audio are much different than the data obtained from any online forums. Therefore, unstructured text data contains much noise, which to some extent can be structured by sentence segmentation. However, sentence segmentation working on only spoken text without punctuation marks or delimiters can cause sentence boundary detection problems. Another challenge in text processing is correcting wrongly transcribed names. For example, "moodle" was wrongly transcribed to "moeder", and we had to manually fix this in the corpus. Therefore, when studies are in-the-wild without a controlled lab environment, then there are more chances for natural, unstructured conversations, which will need cleaning and structuring before analysis can yield meaningful results.

Moreover, the stop word corpus available to the algorithm did not remove all the contextual stop words that were not relevant for this discussion. We also needed to manually remove some contextual stop words like some action verbs depending on their importance in our context by building a contextual stop word library. When we lemmatized and stemmed the words, then the lemmatizer for Dutch text was not accurate enough because of its lesser usage and popularity compared to English. Therefore, we needed to search for local libraries to correct it with some manual intervention.

It is challenging to fully automate the setup. We needed the help of a human to pre-process to some extent for cleaning the corpus, the sanity checks on the names transcribed and to make sense of the visualizations with the help of annotations. Although we are advancing toward automatic collaboration analytics, we are still in an advanced semi-automated phase and need to reduce the dependence on humans in the future.

When constructing the network graph, we quickly run into hairball problems when the graph is filled with many nodes and edges with time. It becomes very difficult to clearly distinguish individual nodes. This can be addressed while designing in the future particularly by using temporal sliders and showing the relevant contextual keywords or words that co-occur above a certain range.

8 Discussion and Conclusion

The literature review gives an overview of unobtrusive measures of collaboration quality and helps to define the quality of collaboration as an event-process conceptual framework. Here, indicators are the events and the indexes which are obtained by processing and aggregating the indicators can be considered as the process. The indicators of collaboration quality are dependent on the scenario of collaboration because of different collaboration task goals and group characteristics (or parameters). Thus, before starting a collaboration task, it should be very clear what are the task goals, what someone wants to measure and how. This is very essential and often overlooked before starting the collaboration task. This can make the prototyping, analytics, and visualization much easier later.

Measuring the collaboration task is complex and needs operationalization of the indicators and indexes of collaboration quality. There needs to be more operationalization of the theoretical indexes into practice. This can help other researchers who want to measure the collaboration quality. For example, there has been a lot of work on measuring "sustaining mutual understanding" with human observers but there has been no work with unobtrusive sensory measures (Praharaj et al., 2021a). It is because of the contextual nuances and difficulty in understanding the content of the conversation which indicates mutual understanding from audio.

Nevertheless, the automated collaboration analytics is in an advanced semi-automated stage and humans are needed to clean the text corpus partially and also correct some names in the transcription. Therefore, there is a need to use good-quality transcription software and contextual keyword corpus to minimize the human dependence and increase the accuracy.

We find that specific keywords utterance frequency analysis for different roles helps to understand the change in role-based conversation patterns with time. This is because the more utterances we have in a specific phase-related keyword, the more is its usage in that context and hence, more importance. The convergence patterns help us to understand how specific conversations were discussed by all roles or specific roles hence signaling an increase in the shared knowledge space (i.e., a proxy for the quality of collaboration). Combined with the social space analysis (shown as role-role interaction network graph), the holistic overview of how the conversations evolved can be obtained. This helped us to quantify the collaboration quality. So, we do not categorize whether higher or lower convergence is good or bad. We just show an approach to quantify collaboration and categorizing is up to the context of collaboration. For instance, in our study, if there is higher convergence for on-topic conversations then it is good for the quality of collaboration but higher convergence for off-topic conversations is bad for collaboration quality. As we do not define fixed objectives before collaboration and do not conduct a lab-based study, so it is quite open to interpretation.

The combined social and epistemic space also helps to clear ambiguity in certain situations when a specific indicator does not give a clear indication about the quality of collaboration. For instance, higher turn-taking signals an increase in collaboration quality only when it is happening on task-related discussion and not on clearing confusion and clarifying about the collaborative task (Kim et al., 2015). This is clear from the epistemic space or in other words the content of the conversation. So, there is a need to do a focus shift to the epistemic space from the social space and both need to be seen side by side to get a holistic overview of who spoke "what" and "how" with whom. Audio in this sense provides a richer picture of collaboration quality in an *unobtrusive* manner. With the rise of privacy and ethical concerns, anonymized audio data can be considered a good unobtrusive measure to detect collaboration quality.

Besides, there needs to be a stakeholder participatory design where their design considerations are taken into account when designing the dashboards to increase its adoption and usage. This is essential when visualizations need to be conveyed as a story on the dashboard and data storytelling can change the narrative of collaboration quality interpretation.

To conclude, our contribution is threefold: (1) to give an overview of the unobtrusive measures of collaboration where we define the quality of collaboration, (2) to build an automatic collaboration analytics setup using the audio data, and (3) to analyze and visualize the collaboration indicators from group audio data to move toward detecting CC quality.

Acknowledgments We would like to thank our colleague, Marcel Schmitz for his contributions in building this game connecting learning analytics with learning design and helping with the data collection and analysis.

References

Bachour, K., Kaplan, F., & Dillenbourg, P. (2010). An interactive table for supporting participation balance in face-to-face collaborative learning. *IEEE Transactions on Learning Technologies, 3*(3), 203–213. https://doi.org/10.1109/TLT.2010.18

Bergstrom, T., & Karahalios, K. (2007). Conversation clock: Visualizing audio patterns in co-located groups. In *40th annual Hawaii international conference on system sciences (HICSS'07)* (pp. 78–78). IEEE. https://doi.org/10.1109/HICSS.2007.151

Blikstein, P. (2013). Multimodal learning analytics. In *Proceedings of the third international conference on learning analytics and knowledge* (pp. 102–106). https://doi.org/10.1145/2460296.2460316

Chandrasegaran, S., Bryan, C., Shidara, H., Chuang, T. Y., & Ma, K. L. (2019). Talktraces: Real-time capture and visualization of verbal content in meetings. In *Proceedings of the 2019 CHI conference on human factors in computing systems* (pp. 1–14). https://doi.org/10.1145/3290605.3300807

Child, S., & Shaw, S. (2015). Collaboration in the twenty-first century: Implications for assessment. *Economics, 21*, 2008.

Dede, C. (2010). Comparing frameworks for twenty-first century skills. *Twenty-first century skills: Rethinking how students learn, 20*(2010), 51–76.

Di Mitri, D., Schneider, J., Specht, M., & Drachsler, H. (2018). From signals to knowledge: A conceptual model for multimodal learning analytics. *Journal of Computer Assisted Learning, 34*(4), 338–349. https://doi.org/10.1111/jcal.12288

Dillenbourg, P. (1999). What do you mean by collaborative learning? In *Collaborative-learning: Cognitive & computational approaches* (pp. 1–19). Elsevier.

Grover, S., Bienkowski, M., Tamrakar, A., Siddiquie, B., Salter, D., & Divakaran, A. (2016). Multimodal analytics to study collaborative problem solving in pair programming. In *Proceedings of the sixth international conference on learning analytics & knowledge* (pp. 516–517). https://doi.org/10.1145/2883851.2883877

Hare, A. P. (1994). Types of roles in small groups: A bit of history and a current perspective. *Small Group Research, 25*(3), 433–448. https://doi.org/10.1177/1046496494253005

Huber, B., Shieber, S., & Gajos, K. Z. (2019). Automatically analyzing brainstorming language behavior with meeter. In *Proceedings of the ACM on human-computer interaction*, 3 (CSCW), 1–17. https://doi.org/10.1145/3359132

Jeong, H., & Chi, M. T. (2007). Knowledge convergence and collaborative learning. *Instructional Science, 35*(4), 287–315. https://doi.org/10.1007/s11251-006-9008-z

Jeong, H., & Hmelo-Silver, C. E. (2010). An overview of CSCL methodologies. In *Proceedings of the ninth international conference on learning sciences (ICLS 2010)* (Vol. 1, pp. 921–928).

Kim, T., Chang, A., Holland, L., & Pentland, A. S. (2008). Meeting mediator: Enhancing group collaboration using sociometric feedback. In *Proceedings of the 2008 ACM conference on computer supported cooperative work* (pp. 457–466). https://doi.org/10.1145/1460563.1460636

Kim, J., Truong, K. P., Charisi, V., Zaga, C., Lohse, M., Heylen, D., & Evers, V. (2015). Vocal turn-taking patterns in groups of children performing collaborative tasks: An exploratory study. In *Sixteenth annual conference of the international speech communication association.*

Kivunja, C. (2015). Exploring the pedagogical meaning and implications of the 4Cs 'super skills' for the twenty-first century through Bruner's 5E lenses of knowledge construction to improve pedagogies of the new learning paradigm. *Creative Education, 6*(02), 224. https://doi.org/10.4236/ce.2015.62021

Lubold, N., & Pon-Barry, H. (2014). Acoustic-prosodic entrainment and rapport in collaborative learning dialogues. In *Proceedings of the 2014 ACM workshop on multimodal learning analytics workshop and grand challenge* (pp. 5–12). https://doi.org/10.1145/2666633.2666635

Meier, A., Spada, H., & Rummel, N. (2007). A rating scheme for assessing the quality of computer-supported collaboration processes. *International Journal of Computer-Supported Collaborative Learning, 2*(1), 63–86. https://doi.org/10.1007/s11412-006-9005-x

Oviatt, S., Hang, K., Zhou, J., & Chen, F. (2015). Spoken interruptions signal productive problem solving and domain expertise in mathematics. In *Proceedings of the 2015 ACM on international conference on multimodal interaction* (pp. 311–318). https://doi.org/10.1145/2818346.2820743

Praharaj, S. (2019). Co-located collaboration analytics. In *Proceedings of the 21st international conference on multimodal interaction* (pp. 473–476). https://doi.org/10.1145/3340555.3356087

Praharaj, S. (2022). *Measuring the unmeasurable?: Towards automatic co-located collaboration analytics*. Doctoral Thesis, Open Universiteit. https://doi.org/10.13140/RG.2.2.18216.65287

Praharaj, S., Scheffel, M., Drachsler, H., & Specht, M. (2018a). MULTIFOCUS: Multimodal learning analytics for co-located collaboration understanding and support. In *Proceedings of the 13th European conference on technology enhanced learning (Doctoral consortium).*

Praharaj, S., Scheffel, M., Drachsler, H., & Specht, M. (2018b). Multimodal analytics for real-time feedback in co-located collaboration. In *European conference on technology enhanced learning* (pp. 187–201). https://doi.org/10.1007/978-3-319-98572-5_15

Praharaj, S., Scheffel, M., Drachsler, H., & Specht, M. (2019). Group coach for co-located collaboration. In *European conference on technology enhanced learning* (pp. 732–736). https://doi.org/10.1007/978-3-030-29736-7_77
Praharaj, S., Scheffel, M., Drachsler, H., & Specht, M. (2021a). Literature review on co-located collaboration modeling using multimodal learning analytics—Can we go the whole nine yards? *IEEE Transactions on Learning Technologies, 14*(3), 367–385. https://doi.org/10.1109/TLT.2021.3097766
Praharaj, S., Scheffel, M., Schmitz, M., Specht, M., & Drachsler, H. (2021b). Towards automatic collaboration analytics for group speech data using learning analytics. *Sensors, 21*(9), 3156. https://doi.org/10.3390/s21093156
Praharaj, S., Scheffel, M., Schmitz, M., Specht, M., & Drachsler, H. (2022). Towards collaborative convergence: Quantifying collaboration quality with automated co-located collaboration analytics. In *Lak22: 12th international learning analytics and knowledge conference* (pp. 358–369). https://doi.org/10.1145/3506860.3506922
Reilly, J. M., Ravenell, M., & Schneider, B. (2018). Exploring collaboration using motion sensors and multi-modal learning analytics. In *Proceedings of International Conference on Educational Data Mining*.
Schneider, B., Sharma, K., Cuendet, S., Zufferey, G., Dillenbourg, P., & Pea, R. D. (2015). 3D tangibles facilitate joint visual attention in dyads. In *Proceedings of the 11th international conference on computer supported collaborative learning* (Vol. 1, pp. 156–165).
Stahl, G., Law, N., & Hesse, F. (2013). Reigniting CSCL flash themes. *International Journal of Computer-Supported Collaborative Learning, 8*(4), 369–374. https://doi.org/10.1007/s11412-013-9185-0
Starr, E. L., Reilly, J. M., & Schneider, B. (2018). Toward using multi-modal learning analytics to support and measure collaboration in co-located dyads. In *Proceedings of the 13th international conference on learning sciences*.
Stiefelhagen, R., & Zhu, J. (2002). Head orientation and gaze direction in meetings. In *CHI'02 extended abstracts on human factors in computing systems* (pp. 858–859). https://doi.org/10.1145/506443.506634
Strijbos, J. W., & Weinberger, A. (2010). Emerging and scripted roles in computer-supported collaborative learning. *Computers in Human Behaviour, 26*(4), 491–494. https://doi.org/10.1016/j.chb.2009.08.006
Tausch, S., Hausen, D., Kosan, I., Raltchev, A., & Hussmann, H. (2014). Groupgarden: Supporting brainstorming through a metaphorical group mirror on table or wall. In *Proceedings of the eighth Nordic conference on human-computer interaction: Fun, fast, foundational* (pp. 541–550). https://doi.org/10.1145/2639189.2639215
Teasley, S., Fischer, F., Dillenbourg, P., Kapur, M., Chi, M., Weinberger, A., & Stegmann, K. (2008). Cognitive convergence in collaborative learning. In *Proceedings of the eighth international conference for the learning sciences – ICLS 2008* (Vol. 3, pp. 360–367).
Terken, J., & Sturm, J. (2010). Multimodal support for social dynamics in co-located meetings. *Personal and Ubiquitous Computing, 14*(8), 703–714. https://doi.org/10.1007/s00779-010-0284-x
Zhou, J., Hang, K., Oviatt, S., Yu, K., & Chen, F. (2014). Combining empirical and machine learning techniques to predict math expertise using pen signal features. In *Proceedings of 2014 ACM workshop on multimodal learning analytics workshop & grand challenge* (pp. 29–36). https://doi.org/10.1145/2666633.2666638

Chapter 7
Unobtrusively Measuring Learning Processes: Where Are We Now?

Shane Dawson

Abstract In this section, we explore how unobtrusive observations can improve our understanding of learning processes. Unobtrusive observation refers to the detection and analysis of aspects of learning extracted or surmised from digital traces of a learner's engagement with technologies. The articles covered in this section delve into various aspects of learning processes, such as self-regulated learning, emotions, motivation, entrepreneurial skills, and problem-solving. Although the topics discussed are diverse, they all centre around a common theme of aligning learner trace data with identified theoretical constructs.

Keywords Learning process · Engagement · Educational technology · Problem-solving · Unobtrusive observation

1 Introduction

For the past decade, the field of learning analytics has faced challenges in accurately aligning trace data, including unobtrusive data sources, with learning processes (Gašević et al., 2015). Learning is a complex phenomenon, and the retrospective analysis of behavioural trace data can provide only limited insights into students' learning processes. Despite the increased adoption of technologies in education, very few studies can empirically demonstrate the impact of learning analytics on student learning (Dawson et al., 2019). Ultimately, the goal of education is to prepare and develop the necessary skills, knowledge, and capabilities of individuals for productive participation in society. This requires a solid foundation in knowledge

S. Dawson (✉)
University of South Australia, Adelaide, Australia
e-mail: Shane.Dawson@unisa.edu.au

V. Kovanovic et al. (eds.), *Unobtrusive Observations of Learning in Digital Environments*, Advances in Analytics for Learning and Teaching,
https://doi.org/10.1007/978-3-031-30992-2_7

and skills, as well as the development of personal and social competencies, such as critical thinking, creativity, and problem-solving. The skills for learning in uncertainty or building sensemaking capabilities are increasingly necessary for future education models. The chapters in this section unpack and highlight the constraints and priorities for future research and allude to new ways of using unobtrusive data to better inform teaching and learning practice. The following section first summarizes the commonalities among the presented works before challenging readers to reflect on missing topics and areas for future discussion. The commentary aims to bring forward perspectives on using unobtrusive data to improve teaching and learning practices.

2 Critical Overview of the Chapters

Collectively, the chapters demonstrate the opportunities afforded by analysing unobtrusive data sources. Chapter 2 by Zheng et al. explores an under-researched area in learning analytics by focusing on measuring emotion dynamics. The research literature demonstrates a clear association between emotion, motivation, and feedback. As such, the chapter delves more deeply into the earlier framing established by Winne. Zheng et al. explain how a learner's emotional state changes over time and in response to the learning context and situation. In short, emotions are not static and fluctuate from moment to moment, from context to context. The authors first present a classification system of emotional dynamics characteristics, namely, emotional variability, instability, inertia, cross-lags, and patterns. In so doing, the authors identify some of the methods for detecting emotions in a non-intrusive way, such as emote aloud, facial and vocal expressions, language and discourse, and physiological sensors. At this point, unobtrusive observation data morph into what could be termed multi-modal data or multi-modal learning analytics.

The ability to effectively make sense of information and solve problems is essential for all learners. However, education has struggled to develop efficient and reliable measures of problem-solving skills, particularly in pedagogical models that involve social dynamics and complex processes. In Chap. 3, Wang et al. report on the use of log data analytics to study problem-solving processes in simulation-based learning environments. The authors examine how features extracted from log data can predict problem-solving outcomes and specific problem-solving practices. The results indicate that the deliberate use of pauses during problem-solving is a crucial feature associated with problem-solving competencies. In this context, the use of pauses as an intentional teacher practice can also be seen to promote metacognition. As framed by Flavell (1979), the concept of metacognition involves both metacognitive knowledge and metacognitive regulation, with the latter involving the ability to manage one's thinking processes. The use of deliberate pauses, along with additional feedback and direction, can support metacognitive regulation and, therefore, the development of problem-solving skills.

In Chap. 4, the authors present a case study to measure leadership skills in a workplace learning context. Assessing complex capabilities or so-called "soft" skills is challenging and context-specific. Most studies to date reporting on the assessment of skills such as leadership tend to employ introspective methods such as self-reported questionnaires and inventories. The chapter presented here clearly details how unobtrusive measures such as learner trace data can offer more scalable and reliable assessments. Interestingly, the authors developed an automated machine learning classifier to extract measures from reflective artefacts incorporated within the learning tasks associated with the case study. This aligns with Winne's call for more information-enriched data to better interpret the learning events for subsequent alignment with theory.

The increased adoption of education technologies has led to an expanse of research mining user interactions to predict learner outcomes, attrition or SRL skills. In Chap. 5, Choi et al. examine the opportunities and challenges in measuring motivational constructs using trace data. The authors draw on the COPES (Conditions, Operations, Products, Evaluations, and Standards) model and how trace data can inform how learners engage in multiple cycles of SRL events. Here, the authors note how clickstream data can be used to understand goal changes over time and identify the external conditions preceding a change in a learner's motivational state. As flagged by Choi and Winne, there is a lack of prior work seeking to produce valid measures of motivation in learning analytics. The authors suggest using the Evidence-Centred Design (ECD) framework to identify a construct's critical attributes and their operational definitions. The article provides an example of using the ECD design pattern to distinguish between performance-oriented and mastery-oriented goals.

Finally, in Chap. 6, Winne presents prior work on the COPES (Conditions, Operations, Products, Evaluations, and Standards) model of SRL to illustrate how underlying information can bring meaning to the learning events and operations students undertake. Winne argues that the inclusion of information-enriched data can better support the interpretation of specific learning events. In essence, Winne posits that understanding or supporting the development of SRL requires information or knowledge of the discrete tasks and standards presented to learners. While this is only a partial component of the overall story, it is integral for aligning trace data with SRL processes. In Winne's terms: "Information-enriched data lend meaning to learning events beyond whether an event occurred".

There are many similarities and alignments in the presented chapters. All chapters cover the relationship between user behaviour and learning intention, from identifying emotional states associated with learning activities to identifying problem-solving skills or complex capabilities. The use of unobtrusive observation data in education calls for greater interdisciplinary research. All chapters reflect this interdisciplinarity. The chapters also highlight the inadequacies, or at least the limitations, of current research methods.

3 What Is Currently Missing in the Modelling of Learning Processes?

There are many strengths to the presented chapters in this section, and the following is by no means intended as a critique of the presented works. More so, the commentary is a reflection on the areas that can be used to complement and extend the current suite of chapters.

As detailed in all the chapters presented in this section, the practice of education has undergone significant change over the past decade. The recent introduction of generative artificial intelligence into education signals the potential for further disruption. Despite changes in the delivery of education, technology adoption, or the need to assess complex capabilities, the importance of feedback remains a consistent theme for supporting student learning. Contemporary research has shifted conceptions of feedback from that of a product to a process (Winstone et al., 2017). While all chapters demonstrate the affordances of unobtrusive observations to measure and support student learning, how such data can also support student agency and feedback remains a challenge. The provision of supportive feedback should be seen as a dialogic process that can develop student feedback literacy.

Unobtrusive data are commonly used for developing student- and teacher-facing learning analytics dashboards to support the development of SRL. As Valle et al. (2021) demonstrated in their systematic review of Learning Analytics Dashboards (LADs), there remains a lack of alignment between stated evaluation measures and target outcomes. Similarly, Matcha et al. (2020) undertook a systematic literature review of learning analytics dashboards (LADs) to determine the impact on learning and teaching. The results indicated that existing LADs are not grounded in learning theory, do not support metacognition, do not provide information about effective learning tactics and strategies, and have limitations in their evaluation.

While there are clearly opportunities to bring unobtrusive data sources into line with feedback, there is much work on how the "pipeline" from course outcomes to design, learning activities, assessment and feedback collectively inter-relate. For instance, Zamecnik et al. (2022) developed a LAD to support collaborative learning and explore how student teams interact and engage with the provided information. The study showed significant diversity in how team members interact with the information depending on their allocated roles. For example, team leaders were noted to be more engaged with data that monitored team collaboration. In this regard, the actual LAD design reflects more event-level information for students and the gap between presented data and intended outcomes is very close. In contrast, many LADs present a significant gap between discrete engagement behaviours and understanding of individual learning progress. While LADs can help teams self-regulate, and instructors can monitor team behaviours, there is a need for further research to investigate student understanding of their learning data and how this can be used for developing feedback literacy. This challenging space was not extensively covered in the presented chapters and is one significant area for future work.

Unobtrusive observations have a rich history in the field of Intelligent Tutoring Systems (ITS). In short, ITS are computer-based systems that provide adaptive learning for students in specific knowledge domains. The goal of ITS is to support learning progress that is tailored to each student's unique strengths and weaknesses. Shute's (2011) concept of stealth assessment was spawned from work in ITS and involves using data generated from students' interactions with digital learning environments to assess their knowledge, skills, and abilities. The concept of stealth here is analogous to unobtrusive observations. Importantly, as framed by Shute (2011) and in the preceding chapters, the goal in analysing these forms of naturally occurring learner data is to increase the frequency and opportunities for formative feedback.

References

Dawson, S., Joksimovic, S., Poquet, O., & Siemens, G. (2019). Increasing the impact of learning analytics. In *Proceedings of the 9th international conference on learning analytics & knowledge* (pp. 446–455). https://doi.org/10.1145/3303772.3303784

Flavell, J. H. (1979). Metacognition and cognitive monitoring: A new area of cognitive–developmental inquiry. *American Psychologist, 34*(10), 906–911. https://doi.org/10.1037/0003-066X.34.10.906

Gašević, D., Dawson, S., & Siemens, G. (2015). Let's not forget: Learning analytics are about learning. *TechTrends, 59*(1), 64–71. https://doi.org/10.1007/s11528-014-0822-x

Matcha, W., Uzir, N. A., Gašević, D., & Pardo, A. (2020). A systematic review of empirical studies on learning analytics dashboards: A self-regulated learning perspective. *IEEE Transactions on Learning Technologies, 13*(2), 226–245. https://doi.org/10.1109/TLT.2019.2916802

Shute, V. J. (2011). Stealth assessment in computer-based games to support learning. In *Computer games and instruction* (pp. 503–524). IAP Information Age Publishing.

Valle, N., Antonenko, P., Dawson, K., & Huggins-Manley, A. C. (2021). Staying on target: A systematic literature review on learner-facing learning analytics dashboards. *British Journal of Educational Technology, 52*(4), 1724–1748. https://doi.org/10.1111/bjet.13089

Winstone, N. E., Nash, R. A., Parker, M., & Rowntree, J. (2017). Supporting learners' agentic engagement with feedback: A systematic review and a taxonomy of Recipience processes. *Educational Psychologist, 52*(1), 17–37. https://doi.org/10.1080/00461520.2016.1207538

Zamecnik, A., Kovanović, V., Grossmann, G., Joksimović, S., Jolliffe, G., Gibson, D., & Pardo, A. (2022). Team interactions with learning analytics dashboards. *Computers & Education, 185*, 104514. https://doi.org/10.1016/j.compedu.2022.104514

Part II
Learning Data

Chapter 8
Data for Unobtrusive Observations of Learning: From Trace Data to Multimodal Data

Vitomir Kovanovic, Roger Azevedo, David C. Gibson, and Dirk Ifenthaler

Abstract In this section, we collected articles that discuss the data needed to unobtrusively observe student learning. While a wide range of data can be utilized for this purpose, the proliferation of digital learning technologies provided many opportunities to collect data from students' interactions with digital learning tools and platforms. Typically referred to as trace or log data, it allows observing students as they learn in real-world learning contexts. As they are usually a byproduct of students' use of digital tools, they require little to no additional effort to be collected. The use of such data is an underlying fuel behind much of the research within the learning analytics field. It allows for quick collection and examination of learning data from a large number of students, providing insights into student learning that were not possible before through more traditional data collection procedures.

Keywords Trace data · Multimodal data · Unobtrusive observation

V. Kovanovic (✉)
Centre for Change and Complexity, University of South Australia, Adelaide, SA, Australia
e-mail: Vitomir.Kovanovic@unisa.edu.au

R. Azevedo
School of Modeling Simulation and Training, University of Central Florida, Orlando, FL, USA
e-mail: roger.azevedo@ucf.edu

D. C. Gibson
Data Science in Higher Education Learning and Teaching, Curtin University, Bentley, WA, Australia
e-mail: david.c.gibson@curtin.edu.au

D. Ifenthaler
Data Science in Higher Education Learning and Teaching, Curtin University, Bentley, WA, Australia

Learning, Design and Technology, University of Mannheim, Mannheim, BW, Germany
e-mail: dirk@ifenthaler.info

V. Kovanovic et al. (eds.), *Unobtrusive Observations of Learning in Digital Environments*, Advances in Analytics for Learning and Teaching,
https://doi.org/10.1007/978-3-031-30992-2_8

1 Section Overview

While trace data is highly valuable in providing unobtrusive insights into student learning, there are significant limitations of such data, primarily due to a lack of context and rich descriptions of students as they engage in learning activities. To address these challenges, the use of multimodal data has witnessed significant interest from researchers. Such data typically involves several *channels*, each providing data and insights about different aspects of student learning. Some of those data include audio and video recordings of students in their learning environments, electrodermal activity recorders, and eye-tracking devices, to name a few. Combining these different channels makes it possible to paint a much richer picture of student learning than possible with simple trace data.

In this section, each chapter focuses on the effective use of multimodal data for understanding one particular aspect of student learning. Those include understanding student engagement, affect and emotions, self-regulation and co-regulation of learning, and student collaboration. As each of these aspects has been extensively covered in the existing research literature, the chapters outline how using different types of data sources moves the state-of-the-art in the unobtrusive measurement of learning. The questions in each chapter further our understanding of the interplay between different aspects of student learning. The following brief introductions provide a quick glimpse of how these authors view the unobtrusive observation of learning processes.

Chapter 9 Fatemeh Salehian Kia, Matthew L. Bernacki, Jeffery A. Greene. *Measuring and Validating Assumptions about Self-Regulated Learning with Multimodal Data*

In their chapter, Salehian Kia et al. provide detailed descriptions of how observational and self-reported data can be collected to provide more comprehensive descriptions of students' self-regulated learning. Using two empirical examples, the authors show strategies and approaches for aligning and mapping observational and self-reported data to provide richer insights into self-regulated learning than possible with only one data source.

Chapter 10 Megan Wiedbusch, Daryn Dever, Shan Li, Mary Jean Amon, Susanne Lajoie, Roger Azevedo. *Measuring Multidimensional Facets of SRL Engagement with Multimodal Data*

To provide valid and reliable insights into student learning, there is a strong need for theoretically grounding data collection, measurement and analysis, with the critical construct in this regard being student engagement. In their chapter, Wieldbusch et al. propose a new theoretical model that captures cognitive, emotional, and behavioral facets of engagement within self-regulated learning. The authors also review current approaches for conceptualizing and measuring student engagement and ways in which multimodal data can advance our understanding of student engagement.

Chapter 11 Philip H. Winne. *Roles For Information In Trace Data Used To Model Self-Regulated Learning*

The chapter of Winne discusses the importance of understanding the information associated with different learning trace data events and its use in understanding students' self-regulated learning. The chapter discusses how different operations and processes manipulate information and how effective understanding of students' cognition, metacognition and motivation requires taking into the account both trace data and the information processed by this trace data.

Chapter 12 Jonna Malmberg, Eetu Haataja, Tiina Törmänen, Hanna Järvenoja, Kateryna Zabolotna, Sanna Järvelä. *Multimodal Measures Characterizing Collaborative Groups' Interaction and Engagement in Learning*

The chapter by Malmberg et al. on how multimodal data can be unobtrusively used to evaluate and gain insights into students' collaborative learning and team collaboration, with a particular focus on cognitive and socio-emotional student interactions and co-regulation of learning and team synchrony. Using an example involving EDA wearables and video recordings of student collaboration, the authors showcase how different types of interaction unfold over time alongside team synchrony, which is measured by the similarity of physiological EDA measures of individual team members.

Chapter 13 Victor Lee. *Electrodermal Activity Wearables and Wearable Cameras as Unobtrusive Observation Devices in Makerspaces*

This chapter describes how data from wrist-wearable devices that capture skin conductance levels, also known as electrodermal activity (EDA), can be used to unobtrusively measure student learning and engagement. Mainly focusing on learning within the makerspace context, Lee provides an overview of the makerspace use in education, the history of electrodermal device use in learning science research and the theoretical construct of engagement, which over time moved from simple attendance to measures of electrodermal activity.

Overall, the data discussed in the chapter involve survey data, trace data, EDA wearable data, and audio and video recordings of student learning. Such data provides detailed and unobtrusive descriptions of student learning, allowing for bringing research closer to real-world learning environments. We hope that you find the following chapters useful and informative and help you further advance your own research involving the unobtrusive measurement of student learning.

Chapter 9
Measuring and Validating Assumptions About Self-Regulated Learning with Multimodal Data

Fatemeh Salehian Kia, Mathew L. Bernacki, and Jeffrey A. Greene

Abstract Individuals who engage in self-regulated learning (SRL) tend to perform better in complex learning tasks. However, learners' ability to self-regulate can vary. To understand and support learners' SRL, collecting information about their engagement in specific learning processes in the context of learning tasks is necessary. However, SRL is sufficiently complex that it is not directly observable. Capturing the SRL processes that occur during learning, as students interact with elements of tasks hosted on virtual learning technologies (e.g., learning management systems; LMS), is possible because learners' actions generate observable events that these technologies log. However, discerning how these events reflect SRL processes poses several major theoretical, methodical, and analytical challenges. To address these challenges, we present two projects to illustrate how researchers validated inferences about SRL processes. We demonstrate how observational indicators drawn from multiple channels of event data must be (a) collected from the technologies' log files and the record of learners' self-reports of their learning process, (b) instrumented to describe learner, event, and context, and (c) integrated and temporally aligned. Afterward, we show how researchers can hypothesize about the SRL processes digital events reflect and test inferences using secondary channels of explanatory data provided by learners during the tasks.

Keywords Self-regulated learning · Multimodal data · Measurement validation

The original version of this chapter was revised. The correction to this chapter is available at https://doi.org/10.1007/978-3-031-30992-2_15

F. Salehian Kia (✉)
University of British Columbia, Vancouver, BC, Canada
e-mail: fatemeh.salehiankia@ubc.ca

M. L. Bernacki · J. A. Greene
University of North Carolina at Chapel Hill, Chapel Hill, NC, USA
e-mail: mlb@unc.edu; jagreene@email.unc.edu

V. Kovanovic et al. (eds.), *Unobtrusive Observations of Learning in Digital Environments*, Advances in Analytics for Learning and Teaching,
https://doi.org/10.1007/978-3-031-30992-2_9

1 Introduction

In recent years, there have been tremendous advances in the theory, research, and practice of self-regulated learning (SRL; i.e., how learners activate and sustain learning goal pursuit via cognitive, motivational, behavioral, and affective processing; Schunk & Greene, 2018). In particular, advances in technology offer opportunities to study SRL unobtrusively. There are different perspectives on how people self-regulate their learning. Complex SRL theoretical frameworks proposed by Zimmerman (2000, 2011), Winne and Hadwin (1998), and Efklides (2011) share assumptions that learning begins with an initial interaction between the assets and dispositions learners bring to the tasks and the opportunities for learning the task affords and constraints it imposes. Further, SRL proceeds in a loosely ordered, cyclical, and iterative process, involving dynamic relationships between motivation, cognition, and emotion, and those relations are often contingent on both those that precede them and the emerging context that prior engagement has produced (Ben-Eliyahu & Bernacki, 2015; Winne & Hadwin, 1998). Over time, the complexity of the assumptions proposed by SRL theorists and which researchers aim to study has warranted the development of a broad and increasingly complex methodological toolkit. Accordingly, researchers have been focused on improving their understanding of SRL through new and promising methods of measuring SRL processes (Azevedo et al., 2017) to study how these complex phenomena unfold during learning tasks.

In the earliest years, SRL was measured by self-reported questionnaires that implicitly positioned SRL as a static phenomenon (Pintrich, 1995). Since then, educational technologies have provided opportunities to track learning activities unobtrusively (Greene & Azevedo, 2010) as dynamic processes that unfold over time (e.g., Winne & Hadwin, 1998), and researchers continue to develop more elaborate methods to capture these phenomena. Despite these recent methodological advances in developing SRL measures that have further expanded researchers' understanding of SRL, the methods applied to develop SRL measures need to be tested for the extent they afford valid inferences of SRL processes, and the conditions that contribute to them.

There are four purposes to this chapter. We (1) introduce the theoretical framework that describes SRL, with a specific focus on contributions made through a decade of learning analytics (LA) research involving unobtrusive measurement methods to study SRL. We consider (2) self-reported and (3) observational methods used to study SRL, and thereafter, (4) introduce the next wave of research involving multimodal designs that can converge multiple methods of data collection to capture many channels of SRL processes and illustrate two cases where overlapping data streams on the same SRL processes enables SRL researchers to observe their co-occurrence and confirm the validity of inferences that can be drawn from one channel, based on corroboration from another.

2 The Theory of Self-Regulated Learning

When people face a complex learning task, they may engage in SRL, a multi-faceted process that includes setting goals and planning the strategic engagement to achieve those goals, enactment of strategies, and then monitoring their approach and adapting based on such evaluations of their performance and products until a goal is attained (Zimmerman & Schunk, 2013). There are multiple well-established models of SRL (Panadero, 2017). Each model offers an alternative perspective on SRL. As Winne (2013) stated, the variety of SRL models does not lessen their validity or decrease their utility; it reveals variance in SRL features. However, all these models have characterized SRL as loosely cyclical processes that a person engages in to perform a complex task that requires monitoring cognition, motivation, affect, and behavior. One of the well-established models of SRL was proposed by Winne and Hadwin (1998), who based their model on an information-processing perspective. We use their model as a reference in this chapter because SRL is conceptualized in it as a series of events that span over time, and such modeling affords opportunities to observe SRL processes as indicated by digital traces of students' interactions within technology learning environments, i.e., click events that span over time (Gašević et al., 2015). Winne and Hadwin's (1998) model includes an SRL cycle of four phases that describe academic task engagement: task definition, planning and goal setting, enactment of tactics and strategies, and adapting. In each of these phases, a learner engages in information processing involving the consideration of *conditions* that guide learning, the selection and engagement of *operations*, and the consideration of *products* of learnings via *evaluations* against the *standards* that describe the learner's goal for task engagement (COPES) (Winne, 2018). In the task definition phase, the learner processes information about the conditions of a task. Next, in the planning and goal setting phase, they set the goals and plan to reach them. The goals are multivariate profiles of standards. The third phase involves the enactment of tactics. A learner performs a task by applying tactics and strategies identified in the previous phase. The fourth phase is an optional, yet pivotal phase of SRL. In this metacognitive adaptation phase, learners make major adaptions in three ways, i.e., deleting conditions under which the operations are carried out, tuning conditions that articulate tactics, or restructuring cognitive conditions to create a new approach.

Educational researchers have studied SRL processes primarily by collecting learners' self-reports about their learning tendencies toward and recollections about learning processes (Wolters & Won, 2017). However, the increasing use of technology across all educational sectors has provided opportunities to observe SRL processes by considering the unobtrusive traces of learner's actions when they engage with digital learning tasks (Greene & Azevedo, 2010), and use these *learning events* (Bernacki, 2018) to test hypotheses that derive from SRL frameworks. These opportunities have changed research on SRL as unobtrusive trace data have been more frequently collected as the internet has become a broadly available medium and a greater proportion of learning happens online. The early data were primarily derived from logs of intelligent tutoring systems (ITSs) in schools (Koedinger et al., 1997), hypertext, and hypermedia systems (Azevedo & Cromley, 2004; Salmerón et al., 2006), and science simulations (Biswas et al., 2005). The increasing use of emerging learning technologies such as massive open online courses (MOOCs) and learning management systems (LMSs) have provided unobtrusive data collection opportunities both in formal and informal learning contexts, which allow researchers to study SRL in authentic learning settings on a large scale.

3 Self-Reported SRL Measurement

The traditional form of collecting SRL data is a self-report wherein individuals respond to prompts regarding their attitudes, behaviors, beliefs, abilities, or knowledge. Self-report includes a broad range of methods such as interviews, questionnaires, and think-aloud protocols (Winne & Perry, 2000). Self-report questionnaires have been the most common type of measure to assess SRL; it is considered to be an "offline" form of measurement because the responses are not collected while students perform a task (Veenman, 2011; Schellings, 2011). However, the advantages of self-report questionnaires are many and appeal to researchers who study SRL. Among them, self-reports are typically low-cost to create, collect, and analyze for a large number of students. Prominent examples of questionnaires used in SRL research are the Motivated Strategies for Learning Questionnaire (MSLQ; Pintrich et al., 1993), the Self-Regulation Strategy Inventory – Self-Report (Cleary et al., 2006), and the Regulation of Learning Questionnaire (McCardle & Hadwin, 2015).

Despite the popularity of questionnaires, there are concerns about students' ability to provide valid reports of their SRL processes. Research on SRL measurement has shown that learners can be inaccurate in reporting their learning behaviors (Rovers et al., 2019; Zhou & Winne, 2012). There is contention among some researchers (Winne & Jamieson-Noel, 2002) that students cannot recall their learning processes with precision and thus cannot serve as accurate reporters of their behaviors. This, therefore, raises questions about the validity of self-reported measures. On the other hand, other researchers, e.g., Karabenick and Zusho (2015), emphasize the importance of understanding that students' self-reports represent students' perception of themselves as a learner, which also contributes to the ways they engage in learning. Thus, claims about validity should be made based on the instrument, the specific circumstances, and purposes in which the data are collected, and for which the instrument is used (Wolters & Won, 2017).

Unlike the self-report questionnaires that ask participants to reflect and summarize their typical SRL process or recount their retrospective engagement in SRL, think-aloud protocols (TAPs) provide a concurrent collection of students' reports about their learning as it occurs. TAPs involve verbal reports where participants describe what they are thinking and doing throughout their engagement in a task (Greene et al., 2017). In most think-aloud studies (Greene et al., 2017) researchers transcribe recordings after sessions, then code students' verbalizations and actions according to a codebook that classifies statements as indicative of macro-level strategies including planning, monitoring, enactment of strategies, and observations about the features of the task or their motivation for engaging during learning. The same verbalizations are coded at the micro-level to reflect more refined traces of SRL processes. Researchers have argued that concurrent reports provide the most accurate data about cognition (Ericsson & Simon, 1984; Fox, 2009). However, there are concerns about asking participants to engage in introspection (i.e., reflection on one's mental processes), suggesting that this might change the nature of their thinking (i.e., reactivity) and challenge the accuracy of such introspection (Greene et al., 2011). Fox et al. (2011) conducted a review of studies using TAPs to investigate the impact of reactivity. Their findings indicated no significant impacts on performance or cognition beyond a slight increase in time needed to complete the task. However, Schooler (2011) raised concerns regarding aspects of non-conscious thought that may be overlooked by this procedure.

Microanalytic protocols are a mixed collection of self-report measures that include both self-reporting via questionnaire and online reporting during learning. In microanalytic protocols, researchers collect students' self-reported task-specific beliefs and SRL strategies (Cleary & Callan, 2018). The contextualized assessments (e.g., utilizing structured interviews) are carried out at different points of an authentic learning activity. This method merges SRL theory and the task characteristics to identify the points in time to administer often open-ended questions before, during, and after a learning activity (Cleary et al., 2012). For instance, Callan et al. (2021) applied microanalytic protocols to examine relations between SRL strategies and performance outcomes in a creative problem-solving task. They used structured interviews at multiple points before, during, and after the problem-solving task measuring students' self-efficacy, strategic planning, strategy use, and self-evaluation. The findings indicated that these SRL processes are related differently to creative assessment outcomes, i.e., fluency, flexibility, originality, and usefulness of solutions generated by fifth and sixth graders.

4 Observational SRL Measurement

Traditional methods of measuring self-regulatory processes, such as self-reports and other offline measures, are limited in their ability to represent the temporal nature of SRL (Roll & Winne, 2015). In contrast, observational data obtained through the learners' use of learning technologies can be used to study temporal features of SRL processes (Azevedo et al., 2018). Two kinds of temporal features are often used to conceptualize learning constructs over time. The first relates to how an individual construct occurs in time and changes over time (Fiel et al., 2018). A second temporal feature is a temporal order in which multiple events occur and can be observed to influence learning in combination, i.e., how learning events or states are sequentially organized. For example, a study by Segedy et al. (2015) proposed a novel approach, called coherence analysis (CA), to capture the surrounding context of digital events by considering the temporal characteristic of SRL processes in a classroom study with *Betty's Brain* to detect students' problem-solving strategies. They considered two digital events as coherent if the second event was based on information generated by the first event, an example of a contextualization where the nature of one event is influenced by a prior occurrence and implication of a first (Winne & Hadwin, 1998). Their results demonstrated relationships between observed SRL behaviors and students' task performance.

4.1 Multimodal Observation of SRL

Several interdisciplinary researchers have improved the representation of the complex nature of temporally unfolding SRL processes by using multimodal designs, that can combine multiple records of trace data that can include log files, eye-tracking, physiological sensors, and screen recordings of learners' interactions with technologies (e.g., Mudrick et al., 2019). These multimodal data are generated from the different data channels within learning technologies, as well as provided through additional instrumentation that can be used to observe learners in laboratory settings (Ochoa et al., 2017). Together, these data are aligned and combined to provide complementary sources of data that together can better reflect the temporal unfolding of the motivational, cognitive, metacognitive, and affective processes described in SRL frameworks (Järvelä et al., 2019). For instance, Malmberg et al. (2019) collected physiological data, video observations, and facial recognition data in the context of collaborative learning to explore how different sources of data can be used to detect self-regulatory components of students' interactions. In another example, Sonnenberg and Bannert (2019) integrated coded think-aloud and trace data to detect the temporal order of SRL phases using two channels of SRL data.

SRL researchers who use multimodal designs consider the component SRL processes most essential to their research question and assemble multimodal data collection models that can capture the variables necessary to investigate it. This process of identifying SRL variables requires that the researcher consider the defining characteristics of the SRL process: the *grain size* at which they aim to observe and the SRL phenomenon, whether the *temporality* of the events that compose that process is essential to the analysis, and how best to capture task conditions that form the *context* in which learning is to be observed (Bernacki, 2018):

Granularity – SRL can be measured at different grain sizes based on the learning environments in which it is studied. To find the representation of SRL, researchers need to consider the granularity level at which they can observe SRL. The main challenge for researchers is then what individual or combination of learning events can represent an SRL sub-process in a learning environment.

Temporality – The SRL process is inherently temporally bounded, which means individual learning events and any combination of learning events should be understood with consideration of the preceding and subsequent events.

Restructuring of raw data is required to capture temporally bounded SRL sub-processes. Aggregating across multiple traces to represent an SRL sub-process should be informed by the granularity in which SRL is studied.

Contextuality – The final feature of SRL that poses a major measurement challenge is capturing the context of learning events. The individual learning event or combination of learning events can be understood in the context where it occurred. Capturing the context of these learning events in open-ended learning environments such as LMSs is a more difficult task because learners also engage in learning activities outside the learning environment.

4.2 Establishing the Validity of Inferences from Observational Data in Multimodal Designs

When data thought to reflect the SRL process are drawn in large quantities using unobtrusive methods of observation (Greene & Azevedo, 2010), the inferences drawn about what such data reflect about learning require validation. As the use of trace data with learning analytics methods continues to grow, Winne (2020) has refocused researchers on the establishment of validity as an essential process in these research efforts. Establishing content or construct validity is a necessary first step before data can be understood to reflect specific SRL processes and research questions can be addressed using those data.

To this point, the multimodal designs we have described were established so that researchers could use multiple channels of data to gather information about distinct SRL processes and align them to understand how these different data sources represent SRL processes that interleave or interact with one another during learning.

For the purposes of establishing the validity of trace data multimodal designs can be developed to collect evidence of the *same* SRL process using two different data channels, with the hope that the same learning event can be observed on both channels, and the data can corroborate and validate one another and future inferences that one may wish to make about learning. These types of studies are limited in number, but they are essential for researchers to undertake when investigating the use of data drawn from a new medium where the learning processes the captured events might reflect are not yet clear. Once the meaning of those events is validated, inferences can be made from those events that reflect something meaningful about an individual's learning process. Those data can then be submitted to analysis – often from many more learners than can be observed with more obtrusive measurements like think-aloud protocols – with greater statistical power and potential to produce empirical findings that can refine SRL frameworks. Further, these data can have implications for educational practice, where validated traces become a powerful resource for observation and potential intervention, where a traced event can be understood to reflect similar phenomena within the experience of future learners who generate it during learning. After validation of the inferences made about digital traces, researchers can then observe such data at the scale and in the contexts where learning happens in formal learning settings such as university courses, where students make substantial use of digital learning tools.

In the space below, we provide two examples of multimodal research designs and initial studies developed to collect (1) a form of digital trace data that is unobtrusively gathered from many learners as they engage in a typical learning task, and (2) the second channel of information from those same learners that can be used to capture the individual's account of the learning processes they meant to undertake when those digital events occurred. The second channel thus has the potential to describe the first and to provide a concurrent account that validates an assumption about what the first event reflects about the learner's SRL process. Once validity evidence is established, the second channel of information can be removed from the data model in future collections, and the inferences about what the first channel reflects can be sustained.

Example 1: Embedded Periodic Self-Report Prompts to Examine Students' and Designers' Assumptions About Stages of SRL During Learning Tasks
In the first example, the research team designed a multimodal study to develop indicators of SRL phases from log data in an LMS and validate the inferences drawn from the traces by adopting top-down theory-guided "knowledge engineering" methods (Salehian Kia et al., 2021). In doing so, we engineered a solution to produce a second data opportunity that augments the first. We embedded a prompting system in the LMS in a series of quasi-experiments in authentic classrooms over 2 or 3 weeks when the students worked on an assignment. The tasks were complex and required students to self-regulate their learning process to complete. These assignments targeted students' problem-solving or critical thinking and reasoning skills. When visiting the LMS course page, the students were prompted to report what they were doing on the assignment. The answers were logged in terms of four SRL phases in which students engaged over 2–3 weeks.

In developing observational indicators of SRL phases, we examined the underlying patterns in logged learning events from the LMS, adopting SRL theory as a lens to identify which learning event or combination of learning events represented actions reflecting an SRL phase. A key challenge was defining the temporal granularity of the log data as a representation of an SRL phase. We decided to create a session of learning events, where the session was defined as the time frame within which the learning events representing a single SRL phase occurred. A session was defined as 20 min based on the typical time students spent in LMS viewing assignment-related pages. Then, a macro-level sequence, or what we called "a sequence of sessions" was used to test the convergence of observed and self-reported SRL phases.

For instance, when a student downloaded a worked example and revisited assignment instructions in the LMS, we characterized that as the student engaging in the enactment phase (i.e., when learners engage in the cognitive strategies they planned to use to achieve their learning goal). These characterizations of learners' behaviors can be biased through their dependence on a researcher's inferencing about the learning task, how learners should engage in it, and how this aligns with a conceptual model of SRL. SRL phase inferences are derived from logged events, and these interpretations can be considered reasonable when the data are contextualized into a class, organized temporally, and interpreted in light of the course's instructional design and in accordance with theoretical assumptions about SRL that apply to the task the design delivers. However, the representation of an SRL phase cannot be considered valid until these inferences are cross-validated with another channel of SRL data to provide support for these assumptions. By periodically collecting students' self-reported answers about the learning activities that they engaged in and their alignment to an SRL phase, we were able to validate inferences via convergence of observed and self-reported SRL indicators.

This last step – aligning and analyzing two channels reflecting the same SRL process – was challenging, but essential in order to test the extent to which the observed SRL phases converged with self-reported SRL phases. The main challenge in the convergence test was the temporal alignment of two channels of SRL data. Because the timestamps of traced events and sequences do not fully co-occur with prompted self-reports, these data were almost never concurrent and needed to be more precisely aligned. In this study, we applied the nearest neighbor technique to merge two sources of SRL data by the closest timestamps (see Fig. 9.1). A predefined cutoff was set as a session length (i.e., 20 min) to match observed and self-reported SRL phases.

Next, we computed Cohen's weighted kappa to compare the two SRL classifications of sessions, i.e., observed and self-reported SRL tags. Cohen's weighted kappa measures the degree of agreement between the two classifiers (Cohen, 1968; Salkind, 2006). The most weight is given to the highest agreement, and less weight is given to cells with a near-perfect agreement (i.e., partial agreement). Because self-regulated learning is a loose cycle of phases that learners engage in, the weighted kappa better measures the partial agreements. Therefore, the disagreement between observed and self-reported SRL measures should not always be treated equally. For example, a student visited an assignment instruction in the LMS for the

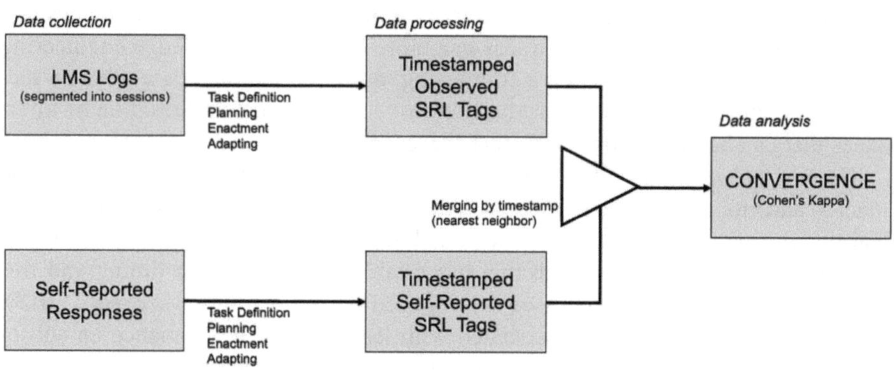

Fig. 9.1 Temporal alignment of self-reported and observed SRL phases

first time to engage in the *task definition* of the learning activity, then the student quickly downloaded the lecture note and worked example files from the LMS course page in close succession to *plan* how to engage in the learning activity. The adjacency of these two phases (i.e., *take definition* and *planning*) in the SRL cycle and close proximal timing of these digital events within one session means that they cannot always be observed individually or separately by an intervening self-report prompt. The prompt asked students to report what they were doing at that particular moment every 20 min if they stayed on the LMS course pages. Thus, the weighted kappa that awards patrial agreement was the closest prompt, in which the student reported their SRL behavior as *planning*, may receive partial agreement when it remains most proximal to prior events that may have represented task definition on the digital trace channel. This disagreement should not be treated as equally important as if a student reported a *task definition* and the observed phase was an *enactment.* Cohen's kappa would treat these two cases equally, whereas the weighted kappa would give more weight to the latter disagreement than the former one.

The results revealed a substantial agreement between the two SRL channels (weighted kappa, $\kappa = .62 - .74$), which were comparably consistent for four student groups in four assignments that targeted either problem-solving or critical thinking skills. When students' SRL was observed and reported in two different courses, one was an introductory computer-programming course and the other was an elective course on theories of game design, students' self-reports of SRL phases corresponded 62–74% (corrected for chance agreement) with the way the researchers believed the digital events should be interpreted. This corroboration across multiple channels strengthens these assumptions and provides confidence that such inference could be validly drawn from future students' trace data, even without asking them to confirm those inferences in future reports.

Example 2: Laboratory Study of Undergraduate Coursework Using Convergent Data from Digital Traces and Verbal Protocols

In the second example, a different research team designed a multimodal study to develop emergent evidence of students' SRL processes using a think-aloud protocol during a lesson completed in an LMS and in the laboratory. They also collected digital traces of learners' concurrent activity in the LMS to validate inferences that can be drawn from these digital traces in a more bottom-up knowledge engineering process. This approach provides a contrasting multimodal design case where a second data opportunity – student verbalizations – can provide confirmation of inferences that can be made from the first (LMS events).

Greene and Azevedo (2007, 2009) have gathered students' accounts about the macro- and micro-level SRL processes they undertake when self-regulating their learning during complex tasks. These coded data have been the medium for a substantial amount of SRL research that has established the relative timing and frequency with which SRL processes are engaged during learning (Azevedo, 2018), the degree to which they are associated with learning and performance on subsequent tasks (Greene et al., 2020), the differences in the learning processes that emerged when tasks differ in their design or the domain of knowledge on which they focus (Greene et al., 2015), and the ways that sequences of events can be gathered to describe adaptive and maladaptive metacognitive monitoring and control processes (Binbasaran Tuysuzoglu & Greene, 2015).

In these prior studies, students' verbalizations are relied upon as a source of information about the learning process and are collected from a relatively small number of learners in a single task designed by researchers. Whereas these data provide a rich description of the learning process and have delivered immense insight into SRL as a phenomenon, they are drawn from small samples in laboratory contexts that differ in many ways from the places where students typically learn. Additionally, the tasks that students complete vary in their similarity to the kinds of assignments that compose the coursework that they complete in formal educational settings as they pursue academic degrees.

In this example work from the *Transformative Undergraduate Self-regulated STEM Learning and Education Research* project, we incorporated think-aloud protocols as one method of data collection within a larger, multimodal effort designed to validate students' digital trace data to the verbal traces that learners traditionally produce in laboratory settings. We worked closely with university science and math instructors to evaluate their course syllabi and the learning tasks students typically encounter in their early undergraduate STEM coursework. We followed these interviews with a lengthy co-design process (Lockyer et al., 2013), wherein the research team observed the instructor's typical lesson, selected a single representative lesson from the end of the semester, and replicated the lesson on a cloned site on the same learning management system, digital video platform, digital textbook, and assessment platform where the students typically complete their coursework. Then, students were recruited from the class to complete this lesson in the laboratory many weeks before they would encounter it in class. In this way, participants in the study would have an opportunity to engage in a task they found authentic to their experience as learners, that had some relevance to their program of study, and that could activate their prior knowledge about how to engage in the complex navigation of the many tools instructors tend to use to provide active learning opportunities in STEM education settings (Lombardi et al., 2021).

The first wave of the project involved a data collection phase where samples of 50–60 students completed laboratory sessions and an initial data analysis phase where their verbal data were transcribed and coded for micro- and macro-level SRL processes, as is typical of think-aloud studies. During this phase, a novel inclusion in the think-aloud data file was a timestamp for each verbalization's onset and end. These timestamps were essential to the second data analysis phase, where those timestamps were aligned to the timestamps of the digital learning events that students initiated when they engaged with the LMS, textbook, video, and assessment platforms.

We obtained the log files of students' activities in each of the laboratory versions of the LMS, textbook, video, and assessment platforms where they learned during the sessions. We prepared these data by aligning their timestamps and the timestamps in the log data, and then classified the objects as they reflected a resource that we inferred should afford a specific SRL process. These included guided reading questions meant to serve as an advanced organizer that frames the reading of a textbook (i.e., a task definition resource, per Greene & Azevedo, 2007 macro codes), textbook passages meant to offer opportunities to acquire information (i.e., micro- and macro-level codes including reading, strategy enactment) and assessment opportunities that provide feedback on one's current knowledge and progress in learning (i.e., information that promotes metacognitive monitoring, and which may precede and inform students' subsequent metacognitive control events involving sustained or adapted strategy use; e.g., Binbasaran Tuysuzoglu & Greene, 2015).

Similar to the first example, developing a method for temporally aligning digital and verbal events proved challenging. To validate our assumptions about the SRL processes students would undertake when using digital resources, we first aligned all verbalizations that had starting and ending timestamps that directly overlapped the instantaneous timestamp that was logged when a student engaged with a digital resource. Thereafter, we expanded the range to capture verbalizations that fell within five, then ten-second windows before or after the digital event. We then examined the degree to which students' verbalizations were homogenously coded as reflecting a single SRL macro process or a single SRL micro process (see Fig. 9.2). If the majority of the students' overlapping or adjacent verbalizations that aligned to a single, specific digitally traced event (e.g., use of RESOURCE) also aligned to a single SRL macro process, we could then infer that the digital event reflected that macro process. Borrowing from standards for internal consistency between multiple items representing a construct and thresholds for adequate inter-rater reliability, we confirmed such inferences about a digital resource when 70% or more of respondents' verbalizations indicated a macro process that described the event of the resources we evaluated in our first study of a biology lesson, these analyses confirmed the majority of our inferences about the SRL macro processes that use of these digital traces of resource use reflect during the lessons students encounter in class. One group of events we could not validate included the digital events students produced when they completed the formative assessments

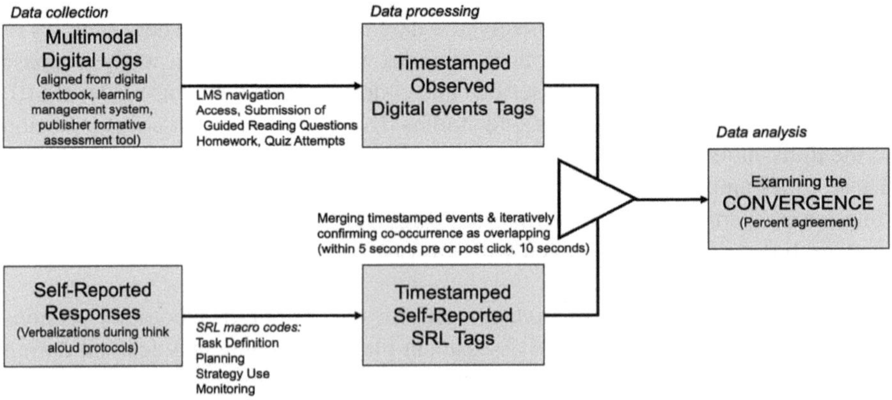

Fig. 9.2 Temporal alignments of digital and verbal channels

embedded in their active learning coursework and the summative assessments they complete after lessons. In these cases, students' verbalizations were only found to be homogenous within individual items, and the nature of their verbalizations was aligned not to the assessment as it reflected an opportunity to learn, but rather as an assessment task where students engaged in test-taking strategies such as ruling out of answers, rather than learning strategies such as making a monitoring judgment in the face of an incorrect attempt and following it up with a revisitation of content where new information could be acquired or one's understanding refined (Bernacki et al., in preparation).

These laboratory validation results have been encouraging, and the second wave of analyses is now underway. In order to establish additional validity about the way these digital events reflect SRL processes, the macro-processes that were assigned to digital events during the lab study are now being applied to data across all the biology lessons that align to the evolution lesson that was sampled into the laboratory as the learning task. Hundreds of students' engagements with that same lesson are being observed in the classroom. We will examine whether the same events occurred in similar quantity and sequence on the sample lesson in this naturalistic environment, then examine whether that pattern extends to the rest of the lessons in the course. Finally, we will aim to establish *predictive* validity by examining whether these SRL processes that are proposed to improve learning and achievement predict variance in students' performance on quizzes and unit exams in the course. These waves of validation allow us to improve our support inferencing from data that can now be collected at the scale of the biology course. Extending the breadth of our observation to classrooms yields considerable statistical power to detect effects and test basic research questions with highly representative samples under authentic classroom conditions. This can enhance our ability to refine theory, and findings can guide the development of future learning resources and the development of learning intervention and support.

4.3 Implications of Multimodal Designs for Research on SRL

These examples of studies aimed at validating SRL inferences to be made from digital trace data using second channels representing the same SRL process illustrated how to validly scale up the use of theory-informed learning analytics in online learning environments. In these two projects, researchers designed distinct instruments for validation. The first example involved a periodic prompting tool embedded in the learning management system and the second example utilized think-aloud protocols with careful logging of timestamps to map with their corresponding inferences about the students' SRL from digital trace data. In the second example, researchers collected think-aloud at a smaller scale and validated the SRL inferences before testing their measures on a larger scale. Both these examples aimed at scaling the validated SRL measures using trace data, and each validation process highlighted the caution when considering the implications of data handling methods during alignment. Mapping the multi-channel data requires highly technical efforts – processing and aligning time stamps, writing rules to select co-occurrences – and theoretical, where inferencing should align to an established schema to describe SRL (i.e., phases, macro-processes).

The opportunity to observe validated SRL events as they occur within many learning activities over the lengthy learning tasks that learners complete affords opportunities to examine complex assumptions about SRL. These include assumptions (1) about the sequential nature of SRL, where certain SRL macro processes should happen in sequence within a single SRL cycle, and where that cycle iterates as students pursue learning goals (2) that task conditions and learner conditions interact to produce contexts where learners may be more or less apt to engage in certain events, or those events may have a context-dependent relationship to future learning or performance, or (3) that events might occur prior to a focal event in one unit and elsewhere in sequence in another, thus providing an opportunity to examine assumptions about contingent relationships (i.e., where the occurrence or importance of an event as a predictor of later learning and performance is conditioned on the contingency where the focal event occurs after or apart from a prior event). In sum, these two examples of cross-validation and scaling efforts provide ways to create opportunities to better understand students' learning processes via careful research on proximal measures of learning in tasks that students will encounter in everyday settings. When those inquiries are aligned with complex assumptions of SRL theory (Ben-Eliyahu & Bernacki, 2015; Bernacki, 2018; Winne & Hadwin, 1998), the results of analyses can further refine such theory.

4.4 Limitations

These two projects demonstrate methods of triangulating channels of behavioral and self-reported data, which can be used to validate one another as measures indicative of SRL processes. Each demonstration comes with limitations. The first limitation is the overfitting of student data to existing taxonomies that describe SRL. In the first case, the provision of a multiple-choice response set to students was an efficient method for enabling them to quickly and clearly indicate their current SRL process. However, providing options rather than an open prompt may have engendered over-reporting of the pre-determined SRL processes and precluded reporting of others students would think to disclose. In the second case, a pre-existing codebook developed to categorize verbalizations during prior think-aloud studies focused on hypermedia was applied to learning in a highly structured active learning lecture. Imposing that codebook constrained coder thinking to the rules that classified SRL processes common to prior tasks, and may have narrowed the expansion of the codebook to accommodate the new task, limiting the SRL micro-process types that could be observed.

The second limitation to this triangulation for validation involves the misalignment by time and grain size of SRL events across channels of data. In the digital channels of both studies, logs are recorded for specific, active events that involve a recordable click, entry, or other action, but not when someone engages in a reflective act that requires no action. This creates a mismatch where coded think-aloud data are voluminous and more diverse in the types of SRL processes they capture. The think-aloud can be used to validate the digital traces, but digital traces are insufficient for validating the subset of SRL processes like some planning and monitoring events that are described, but that have no corresponding action. Responses to multiple-choice reports of SRL processes induce related difficulties where event logs fail to capture the entirety of learners' SRL process on a digital side, where embedding the self-report tool was technically challenging and led to some loss of reporting data, as did outages and technical difficulties with event logs. These infrastructure challenges were further exacerbated by issues with data alignment where self-report prompts can only be issued periodically without becoming a nuisance, and such logs of self-reported SRL processes are necessarily sparse. When learning tasks are dense with required actions, those self-reports are too sparse to map 1-to-1 with actions, and consequently, some self-reports of SRL processes were inferred to apply to a number of digitally observed SRL events tags for which a single self-report was the closest timestamped report in the log. This leads to a looser approximation of the SRL event as described by the SRL report, and this looser alignment can inflate disagreements between self-reported and observed tags. A final limitation is a theoretical one: the SRL measures were developed based on the COPES model from an information-processing perspective (Winne & Hadwin, 1998), which focused coding on some SRL processes key to this conceptualization, and precluded coding of SRL processes key to other models (e.g., activating Metacognitive Knowledge; Efklides, 2011). Additional coding schemes would need to be incorporated to fully map the SRL framework more in future observational or validation studies.

5 Conclusions

In this chapter, we discussed methods for leveraging the affordances of observed and self-report data (e.g., direct observation and solicitation of participants' behaviors and perceptions of those behaviors) to validate the inferences necessary to take advantage of the affordances of unobtrusive trace data collection at scale. The resource-intensive process of collecting multimodal, multichannel data in an authentic environment is well-warranted when it results in the validation of what digital trace data indicate about learners' SRL and thus allows for confident inferences from those data. As shown in this chapter, there are several promising methods for aligning these multimodal, multichannel data, resulting in stronger evidence of the validity of inferences from digital traces. More research is needed to determine the most efficient and effective ways of doing this work, which promises to greatly enhance the field's ability to draw valid, reproducible, and useful inferences about SRL from digital trace data.

References

Azevedo, R. (2018). Using hypermedia as a metacognitive tool for enhancing student learning? The role of self-regulated learning. In *Educational psychologist* (pp. 199–209). Routledge.

Azevedo, R., & Cromley, J. G. (2004). Does training on self-regulated learning facilitate students' learning with hypermedia? *Journal of Educational Psychology, 96*(3), 523.

Azevedo, R., Taub, M., & Mudrick, N. V. (2017). Understanding and reasoning about real-time cognitive, affective, and metacognitive processes to foster self-regulation with advanced learning technologies. In *Handbook of self-regulation of learning and performance* (pp. 254–270). Routledge.

Azevedo, R., Taub, M., Mudrick, N. V., Martin, S. A., & Grafsgaard, J. (2018). Using multichannel trace data to infer and foster self-regulated learning between humans and advanced learning technologies. In *Handbook of self-regulation of learning and performance* (2nd ed., pp. 254–270). Routledge.

Ben-Eliyahu, A., & Bernacki, M. L. (2015). Addressing complexities in self-regulated learning: A focus on contextual factors, contingencies, and dynamic relations. *Metacognition and Learning, 10*(1), 1–13.

Bernacki, M. L. (2018). Examining the cyclical, loosely sequenced, and contingent features of self-regulated learning: Trace data and their analysis. In *Handbook of self-regulation of learning and performance* (pp. 370–387). Routledge.

Binbasaran Tuysuzoglu, B., & Greene, J. A. (2015). An investigation of the role of contingent metacognitive behavior in self-regulated learning. *Metacognition and Learning, 10*(1), 77–98.

Biswas, G., Leelawong, K., Schwartz, D., Vye, N., & The Teachable Agents Group at Vanderbilt. (2005). Learning by teaching: A new agent paradigm for educational software. *Applied Artificial Intelligence, 19*(3–4), 363–392.

Callan, G. L., Rubenstein, L. D., Ridgley, L. M., & McCall, J. R. (2021). Measuring self-regulated learning during creative problem-solving with SRL microanalysis. *Psychology of Aesthetics, Creativity, and the Arts, 15*(1), 136.

Cleary, T. J., Zimmerman, B. J., & Keating, T. (2006). Training physical education students to self-regulate during basketball free throw practice. *Research Quarterly for Exercise and Sport, 77*(2), 251–262.

Cleary, T. J., Callan, G. L., & Zimmerman, B. J. (2012). Assessing self-regulation as a cyclical, context-specific phenomenon: Overview and analysis of SRL microanalytic protocols. *Education Research International, 2012*, 1.

Cleary, T. J., & Callan, G. L. (2018). Assessing self-regulated learning using microanalytic methods. In D. H. Schunk & J. A. Greene (Eds.), Handbook of self-regulation of learning and performance (pp. 338–351). Routledge/Taylor & Francis Group

Cohen, J. (1968). Weighted kappa: nominal scale agreement provision for scaled disagreement or partial credit. *Psychological bulletin, 70*(4), 213.

Efklides, A. (2011). Interactions of metacognition with motivation and affect in self-regulated learning: The MASRL model. *Educational Psychologist, 46*(1), 6–25.

Ericsson, K. A., & Simon, H. A. (1984). Protocol analysis: Verbal reports as data. The MIT Press.

Fiel, J., Lawless, K. A., & Brown, S. W. (2018). Timing matters: Approaches for measuring and visualizing behaviours of timing and spacing of work in self-paced online teacher professional development courses. *Journal of Learning Analytics, 5*(1), 25–40.

Fox, E. (2009). The role of reader characteristics in processing and learning from informational text. *Review of Educational Research, 79*(1), 197–261.

Fox, M. C., Ericsson, K. A., & Best, R. (2011). Do procedures for verbal reporting of thinking have to be reactive? A meta-analysis and recommendations for best reporting methods. *Psychological Bulletin, 137*(2), 316.

Gašević, D., Dawson, S., & Siemens, G. (2015). Let's not forget: Learning analytics are about learning. *TechTrends, 59*(1), 64–71.

Greene, J. A., & Azevedo, R. (2007). A theoretical review of Winne and Hadwin's model of self-regulated learning: New perspectives and directions. *Review of Educational Research, 77*(3), 334–372.

Greene, J. A., & Azevedo, R. (2009). A macro-level analysis of SRL processes and their relations to the acquisition of a sophisticated mental model of a complex system. *Contemporary Educational Psychology, 34*(1), 18–29.

Greene, J. A., & Azevedo, R. (2010). The measurement of learners' self-regulated cognitive and metacognitive processes while using computer-based learning environments. *Educational Psychologist, 45*(4), 203–209.

Greene, J. A., Robertson, J., & Costa, L. J. C. (2011). Assessing self-regulated learning using think-aloud methods. In *Handbook of self-regulation of learning and performance* (pp. 313–328). Routledge.

Greene, J. A., Bolick, C. M., Jackson, W. P., Caprino, A. M., Oswald, C., & McVea, M. (2015). Domain-specificity of self-regulated learning processing in science and history. *Contemporary Educational Psychology, 42*, 111–128.

Greene, J. A., Deekens, V. M., Copeland, D. Z., & Yu, S. (2017). Capturing and modeling self-regulated learning using think-aloud protocols. In *Handbook of self-regulation of learning and performance* (pp. 323–337). Routledge.

Greene, J. A., Lobczowski, N. G., Freed, R., Cartiff, B. M., Demetriou, C., & Panter, A. T. (2020). Effects of a science of learning course on college students' learning with a computer. *American Educational Research Journal, 57*(3), 947–978.

Järvelä, S., Malmberg, J., Haataja, E., Sobocinski, M., & Kirschner, P. A. (2019). What multimodal data can tell us about the students' regulation of their learning process. *Learning and Instruction, 72*(7), 4.

Karabenick, S. A., & Zusho, A. (2015). Examining approaches to research on self-regulated learning: Conceptual and methodological considerations. *Metacognition and Learning, 10*(1), 151–163.

Koedinger, K. R., Anderson, J. R., Hadley, W. H., & Mark, M. A. (1997). Intelligent tutoring goes to school in the big city. *International Journal of Artificial Intelligence in Education, 8*(1), 30–43.

Lockyer, L., Heathcote, E., & Dawson, S. (2013). Informing pedagogical action: Aligning learning analytics with learning design. *American Behavioral Scientist, 57*(10), 1439–1459.

Lombardi, D., Shipley, T. F., & Astronomy Team, Biology Team, Chemistry Team, Engineering Team, Geography Team, Geoscience Team, and Physics Team. (2021). The curious construct of active learning. *Psychological Science in the Public Interest, 22*(1), 8–43.

Malmberg, J., Järvelä, S., Holappa, J., Haataja, E., Huang, X., & Siipo, A. (2019). Going beyond what is visible: What multichannel data can reveal about interaction in the context of collaborative learning? *Computers in Human Behavior, 96*, 235–245.

McCardle, L., & Hadwin, A. F. (2015). Using multiple, contextualized data sources to measure learners' perceptions of their self-regulated learning. *Metacognition and Learning, 10*(1), 43–75.

Mudrick, N. V., Azevedo, R., & Taub, M. (2019). Integrating metacognitive judgments and eye movements using sequential pattern mining to understand processes underlying multimedia learning. *Computers in Human Behavior, 96*, 223–234.

Ochoa, X., Lang, A. C., & Siemens, G. (2017). Multimodal learning analytics. In *The handbook of learning analytics* (1st ed., pp. 129–141). Society for Learning Analytics Research.

Panadero, E. (2017). A review of self-regulated learning: Six models and four directions for research. *Frontiers in Psychology, 8*, 422.

Pintrich, P. R. (1995). Understanding self-regulated learning. *New Directions for Teaching and Learning, 1995*(63), 3–12.

Pintrich, P. R., Smith, D. A., Garcia, T., & McKeachie, W. J. (1993). Reliability and predictive validity of the Motivated Strategies for Learning Questionnaire (MSLQ). *Educational and Psychological Measurement, 53*(3), 801–813.

Roll, I., & Winne, P. H. (2015). Understanding, evaluating, and supporting self-regulated learning using learning analytics. *Journal of Learning Analytics, 2*(1), 7–12.
Rovers, S. F., Clarebout, G., Savelberg, H. H., de Bruin, A. B., & van Merriënboer, J. J. (2019). Granularity matters: Comparing different ways of measuring self-regulated learning. *Metacognition and Learning, 14*(1), 1–19.
Salehian Kia, F., Hatala, M., Baker, R. S., & Teasley, S. D. (2021, April). Measuring students' self-regulatory phases in LMS with behavior and real-time self report. In *LAK21: 11th international learning analytics and knowledge conference* (pp. 259–268).
Salkind, N. J. (2006). *Encyclopedia of measurement and statistics*. Sage.
Salmerón, L., Kintsch, W., & Caãs, J. J. (2006). Reading strategies and prior knowledge in learning from hypertext. *Memory & Cognition, 34*(5), 1157–1171.
Schellings, G. (2011). Applying learning strategy questionnaires: Problems and possibilities. *Metacognition and Learning, 6*(2), 91–109.
Schooler, J. W. (2011). Introspecting in the spirit of William James: Comment on Fox, Ericsson, and Best (2011). *Psychological Bulletin, 137*(2), 345–350.
Schunk, D. H., & Greene, J. A. (Eds.). (2018). Handbook of Self-Regulation of Learning and Performance (2nd ed.). Routledge.
Segedy, J. R., Kinnebrew, J. S., & Biswas, G. (2015, June). Coherence over time: Understanding day-to-day changes in students' open-ended problem-solving behaviors. In *International conference on artificial intelligence in education* (pp. 449–458). Springer.
Sonnenberg, C., & Bannert, M. (2019). Using Process Mining to examine the sustainability of instructional support: How stable are the effects of metacognitive prompting on self-regulatory behavior? *Computers in Human Behavior, 96*, 259–272.
Veenman, M. V. (2011). Alternative assessment of strategy use with self-report instruments: A discussion. *Metacognition and Learning, 6*(2), 205–211.
Winne, P. H. (2013). Self-regulated learning viewed from models of information processing. In *Self-regulated learning and academic achievement* (pp. 145–178). Routledge.
Winne, P. H. (2018). Cognition and metacognition within self-regulation. In *Handbook of self-regulation of learning and performance*. Routledge.
Winne, P. H. (2020). Commentary: A proposed remedy for grievances about self-report methodologies. *Frontline Learning Research, 8*(3), 164–173.
Winne, P. H., & Hadwin, A. F. (1998). Studying as self-regulated learning. In D. J. Hacker, J. Dunlosky, & A. C. Graesser (Eds.), *Metacognition in educational theory and practice* (pp. 277–304). Erlbaum.
Winne, P. H., & Jamieson-Noel, D. (2002). Exploring students' calibration of self reports about study tactics and achievement. *Contemporary Educational Psychology, 27*(4), 551–572.
Winne, P. H., & Perry, N. E. (2000). Measuring self-regulated learning. In *Handbook of self-regulation* (pp. 531–566). Academic.
Wolters, C. A., & Won, S. (2017). Validity and the use of self-report questionnaires to assess self-regulated learning. In *Handbook of self-regulation of learning and performance* (pp. 307–322). Routledge.
Zhou, M., & Winne, P. H. (2012). Modeling academic achievement by self-reported versus traced goal orientation. *Learning and Instruction, 22*(6), 413–419.
Zimmerman, B. J. (2000). Attaining self-regulation: A social cognitive perspective. In *Handbook of self-regulation* (pp. 13–39). Academic.
Zimmerman, B. J., & Schunk, D. H. (2011). *Handbook of self-regulation of learning and performance*. Routledge/Taylor & Francis Group.
Zimmerman, B. J., & Schunk, D. H. (2013). Reflections on theories of self-regulated learning and academic achievement. In *Self-regulated learning and academic achievement* (pp. 282–301). Routledge.

Chapter 10
Measuring Multidimensional Facets of SRL Engagement with Multimodal Data

Megan Wiedbusch, Daryn Dever, Shan Li, Mary Jean Amon, Susanne Lajoie, and Roger Azevedo

Abstract Essential to achieving adaptive intelligent AI-based education systems is theoretically grounded data measurement and analysis, and the subsequent data-supported individualized interventions that foster learner-system engagement. However, engagement is a challenging psychological construct to define and measure given the variation of theoretical conceptualizations of engagement and the various facets of engagements (e.g., behavioral, emotional, agentic, (meta)cognitive, and self-regulated learning). In this chapter we (1) define and situate a multifaceted conceptualization of engagement (based on the interrelated aspects of student engagement) within SRL, (2) introduce the integrative model of multidimensional self-regulated learning engagement to include cognitive, emotional, and behavioral facets of engagement; (3) briefly review the current conceptual, theoretical, and methodological approaches to measuring engagement and showcase how the use of multimodal data for this work has contributed to our understanding of learning in learning systems. Engagement-relevant data discussed within this chapter includes self-reports, log or behavioral streams, oculometrics, physiological sensors (e.g., skin conductance, heart-rate, etc.), facial expressions, body gestures, and think- and emote-alouds. We can leverage these multimodal data to reflect the dynamic and nonlinear nature of engagement that are frequently obfuscated by traditional unimodal methods (e.g., self-reports). However, it is crucial that when multimodal data is converged for this purpose, we consider a unifying theoretical grounding of engagement that is general enough to be applied across intelligent

M. Wiedbusch (✉) · D. Dever · M. J. Amon · R. Azevedo
University of Central Florida, Orlando, FL, USA
e-mail: megan.wiedbusch@ucf.edu; daryn.dever@ucf.edu; mary.jean.amon@ucf.edu; roger.azevedo@ucf.edu

S. Li
Lehigh University, Bethlehem, PA, USA
e-mail: shla22@lehigh.edu

S. Lajoie
McGill University, Montreal, QC, Canada
e-mail: susanne.lajoie@mcgill.ca

V. Kovanovic et al. (eds.), *Unobtrusive Observations of Learning in Digital Environments*, Advances in Analytics for Learning and Teaching,
https://doi.org/10.1007/978-3-031-30992-2_10

systems and the contexts in which they are used but specific enough to be useful in the design and development of analytical methods.; and (4) provide a methodological overview with contextualized examples to inform the research study design of future testing and validation of our integrative model of multidimensional self-regulated learning engagement using multimodal data. Our methodological overview identifies how different modalities of measurement and their temporal granularity contribute to the measurement of engagement as it fluctuates within the different phases of self-regulated learning. We conclude our chapter with an exploration of the implications of this guide as well as future directions for researchers, instructional designers, and software engineers capturing and analyzing engagement in digital environments using multimodal data.

Keywords Engagement · Multimodal data · AI-based education (AIED) systems

1 Introduction

Engagement is not just how involved a learner is with their task, but rather is goal-directed action that serves to help an individual progress academically, satisfy motivations, and create motivationally supportive learning environments (Reeve et al., 2019). Engagement has the potential to tackle persistent educational issues of low achievement (Boekaerts, 2016; Sinatra et al., 2015), risk-behaviors (Reschly & Christenson, 2012; Wang & Fredricks, 2014), and high rate of student boredom and alienation (Chapman et al., 2017; Fredricks et al., 2016, 2019a, b). However, as many researchers draw attention to, there is a notable inconsistency in both the conceptual definition and measurement of engagement (Azevedo, 2015; Fredricks et al., 2019a; Li & Lajoie, 2021). Sinatra et al. (2015) suggest that this lack of clarity derives from the construct of engagement being developed out of the assessment approach instead of grounding engagement research in a theoretical framework explicitly.

In this chapter, we address this issue by extending the Integrative Model of Self-Regulated Learning (SRL) Engagement (Li & Lajoie, 2021) to a multimodal approach for measuring the multidimensional facets of engagement. We begin by broadly defining engagement as a multifaceted construct and its relationship to self-regulated learning (SRL). Next, we introduce the extension of Li and Lajoie's (2021) Integrative Model of SRL Engagement to include additional dimensions of engagement (i.e., cognitive, behavioral, and emotional engagement). This is followed by a brief description of new underlying assumptions, strengths, and challenges of this model. We then review various unimodal methods of measuring engagement as a basis for data channel convergence and multimodal assessment. Additionally, we provide a conceptual approach of our own by providing examples of how various data channels can be used to interpret student engagement using the newly proposed model. We conclude with a discussion of limitations, future directions, and implications for designing and developing AIEd systems that utilize theoretically grounded multimodal measures of engagement.

2 What Is Engagement?

Engagement during learning is a multidimensional construct with four distinct but intercorrelated aspects—behavioral, emotional, cognitive, and agentic—that refers to the extent of a student's active involvement in their learning (Fredricks et al., 2004; Reeve, 2012 extension of Connell & Wellborn, 1991). It is a construct that is inherently dynamic as it ebbs and flows during learning, whether that be for a single task (at the granularity of minutes) or across entire courses (at the granularity of months). When measuring and assessing a learner's quality and quantity of engagement, one must consider the level of attention and effort (behavioral engagement), the depth and quality of the strategy use sophistication (cognitive engagement), presence of facilitating and inhibiting emotions of interest and curiosity (emotional engagement), and the agency with which the learner is able to manipulate and adapt their own learning (agentic engagement). Below we briefly define these four facets.

Behavioral engagement refers to the learner's effortful involvement in their learning through strategy use and activities to stay on task via attention, effort, and persistence (Skinner et al., 2009; Reeve et al., 2019).

Cognitive engagement refers to "the extent to which individuals think strategically along a continuum across the learning or problem-solving process in a specific task" (Li & Lajoie, 2021, p. 2).

Emotional engagement refers to the presence of task-related emotions that may support or inhibit other types of engagement such as interest, curiosity, and anxiety (Reeve, 2013).

Agentic engagement refers to a learner's constructive contribution to their learning such as offering suggestions, asking questions, recommending objectives, and seeking opportunities to steer their learning (Reeve, 2013). In this chapter we do not directly address agentic engagement, however when considering the development of adaptive and intelligent systems, agentic engagement may play a vital role, especially when considering the independence of a learner who is self-regulating.

3 Extension of the Integrative Model of Self-Regulated Learning (SRL) Engagement

Just as engagement is a multidimensional construct, self-regulated learning (SRL) is also a multidimensional construct that refers to the active modulation and regulation of one's learning (see Panadero, 2017 for a review of SRL models). Researchers

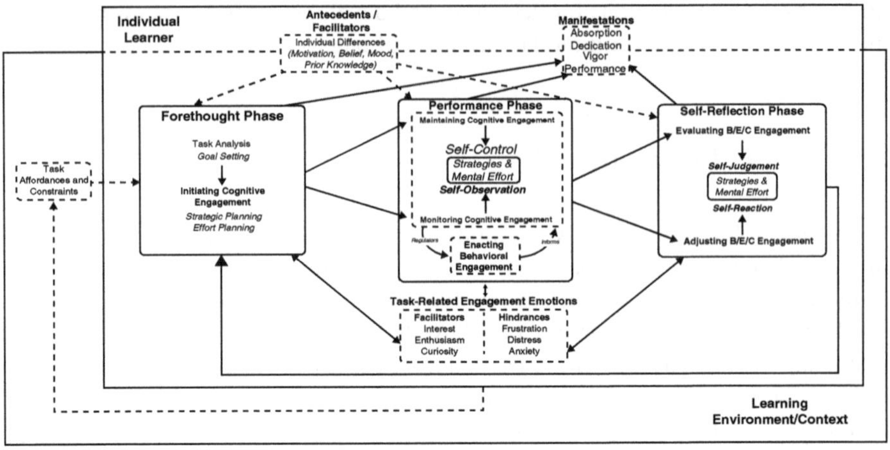

Fig. 10.1 Integrative Model of Multidimensional SRL Engagement (ISSME)

have suggested that due to the large overlap between the two constructs, we should consider integrating them (Wolters & Taylor, 2012), a call that recently Li and Lajoie (2021) responded to with their introduction of the Integrative Model of Self-Regulated Learning Engagement. This new model situates cognitive engagement inside of SRL to improve how we understand how, why, what, and when learners are more efficient and effective learners. Specifically, their model suggests that cognitive engagement fluctuates in both quality and quantity continuously throughout three sequential phases of SRL as described in Zimmerman (2000) – forethought phase (task analysis, goal setting, and strategic planning), performance phase (self-control of task strategy and self-observation), and self-reflection phase (self-evaluation, causal attribution, and adaptive self-reaction). We propose the Integrative Model of Multidimensional SRL Engagement (IMMSE) an extension of this model to include the additional facets of engagement (i.e., behavioral and emotional engagement; see Fig. 10.1).

As an expansion of the original model and its assumptions, the IMMSE places the learner (darker outline) within the learning environment or context (outermost box). Note that part of the learner is situated outside of the learning context (top dotted line). This highlights that the learner brings certain individual differences (e.g., personality, working memory capacity, prior knowledge) from the previous learning experiences such as motivations, beliefs, and moods, which will impact their current and future learning. Individual differences impact the learner both inside and outside of the specific learning context. Additionally, learners take manifestations (e.g., new beliefs, feelings of efficacy, vigor) of their learning out of the current context that will continue to impact their future learning experiences. The IMMSE suggests that those facilitators and manifestations cycle through each phase of SRL but not necessarily in a linear fashion.

In the forethought phase, a learner analyzes their task to set goals. This analysis will then initiate cognitive engagement as a learner plans how best to achieve those goals and with how much effort they should use. We show that task affordances and constraints from the environmental context such as available tools, scaffolding techniques, or pedagogical support must be considered within the forethought phase for goal setting and strategic planning. These are fed in from the learning environment or context into the individual learner. When we consider what engagement-sensitive AIEd systems will look like, we then have to consider the direct effect the learner's behavior, emotions, and cognition will have on the environment. This is depicted as the dotted line running from the leaner back into the task affordances and constraints. We discuss this feedback loop in greater detail in Sect. VI. Limitations and Future Directions.

Next, learners move into the performance phase where they cognitively regulate and monitor their behavioral engagement with the learning environment. The IMMSE argues that cognitive efficiency is a core feature of SRL engagement, such that learners' cognitive monitoring helps inform how best one should strategically and efficiently manage their cognitive engagement (Li & Lajoie, 2021). During the performance phase, cognitive engagement regulates the enactment of behavioral engagement. Behavioral engagement, according to Reeve, Cheon, & Jang, (2019), must be observable and therefore is enacted only within the performance phase. This enactment then informs cognitive engagement maintenance and monitoring mechanisms. As many of the studies that measure engagement demonstrate, cognitive engagement is usually measured by learner use of strategies. The IMMSE further extends from the original integrative model of SRL engagement (Li & Lajoie, 2021) by creating a distinction between engagement and strategy use, a commonly used proxy of cognitive engagement. Our model posits that cognitive engagement helps regulate behavioral engagement, which can help explain when behaviors indicate one level of engagement, but effort and attention indicate another. For example, behaviorally, one may appear to be reading – their eyes scanning over a page. However, when asked about what was just read, the learner may not be able to recall or even mention having their thoughts trail off. In this way, we can show there is behavioral engagement in the use of a strategy, but that there is low cognitive engagement and, in turn, low-quality engagement.

During the self-reflection phase, learners evaluate and adjust their behavioral, emotional, and cognitive engagement based on their reaction and evaluation of those strategies and the effort exerted. This is a slight expansion of the original integrative model such that, not only are learners reflecting on their cognitive engagement, but also on their emotional and behavioral engagement. For example, one might reflect that their interest (emotion engagement) is waning in a particular topic but that they recognize that topic as vital to achieving a particular learning subgoal (cognitive engagement). They determine that their approach at notetaking (behavioral engagement) has caused their declining interest. In the subsequent forethought phase after this reflection, they then plan to update their approach based on this next context and understanding of both themselves, their tasks, and current progress toward their goals.

Another expansion includes the addition of task-related engagement emotions that both facilitate and inhibit the other types of engagement throughout all of the SRL phases. Emotional engagement is unique comparatively to the other two types of engagement as it can inhibit or facilitate the amount of engagement activity that is available to the learner (Reeve et al., 2019). For example, when emotional engagement facilitators, such as an interest, are high, a learner may exert more cognitive effort than normally applied. In the IMMSE, task-related engagement emotions have a bi-directional relationship between each phase. This assumes that these emotions act as both catalysts for the internal cognitive and behavioral processes of SRL and engagement as well as products of those same processes. The IMMSE does not make any explicit assumptions about how those emotions are regulated and modified within each phase (see Harley et al., 2019).

Overall, the IMMSE suggests that engagement is an ever-changing process that fluctuates within learning. However, it is important to note that even if engagement is low quantitatively, this is not the same as disengagement. Many researchers have begun to theorize that engagement and disengagement are two distinct processes that lead to different learning consequences (Cheon et al., 2018; Jang et al., 2016; Haerens et al., 2015; Reeve et al., 2019). As such, this model does not make any distinct assumptions about disengagement but is rather focused on the temporal fluidity of engagement. This fluidity has many interconnected components that are often only ever given a cursory glance in many models of SRL such as emotions and motivations.

The IMMSE provides a theoretical grounding for future research to examine how best to measure engagement, and subsequently use those measurements for adaptive design. Additionally, it provides the groundwork for which researchers specify what it is exactly they are measuring to help clarify some of the conceptual confusion across studies. In the next sections, we review how previous work has measured engagement before providing our own approach that is grounded within this model.

4 Unimodal Methods for Studying Engagement

In this next section, we review how previous research has captured and analyzed engagement while highlighting the strengths and limitations of each approach. We follow by providing some of the new attempts at converging data channels for studying engagement. It is important to note that this review is not exhaustive in nature as some methods such as gesture recognition (e.g., Ashwin & Guddeti, 2019), teacher ratings (e.g., Fredricks & McColskey, 2012), or administrative (or institutional) data (e.g., Mandernach, 2015) have been previously used as measures of engagement but are not discussed for the sake of brevity. This section should serve as a general overview for some of the methods that have been used to measure engagement, but we direct readers to additional conceptual and systematic reviews for additional studies (Azevedo, 2015; Dewan et al., 2019; Fredricks et al., 2016, 2019a, b; Henrie et al., 2015; Li, 2021).

4.1 Clickstream Data/Log Files

Log files are sequential events or data streams where the concurrent interactions of an individual with a system (i.e., human-machine interaction) are captured (Oshima & Hoppe, 2021). Specifically, log-file data typically records the initiator (e.g., student, pedagogical agent) of an action on objects or elements within a system and what time point this action was initiated or completed. Log-file data have been used throughout literature to capture, measure, and analyze student engagement within virtual learning environments throughout a variety of domains and contexts including science (Gobert et al., 2015; Li et al., 2020), education (Henrie et al., 2015), computer science (Shukor et al., 2014), etc.

The use of log-file data can assist researchers in revealing evidence of disengaged behaviors, learner profiles of engagement, and how this relates to learners' overall learning outcomes. For example, a study by Gobert et al. (2015) utilized log files to calculate learners' frequency of actions, the amount of time between actions, and duration of the actions as they learned about ecology with a microworld. Results from this study found that using log files to measure engagement can indicate gaming the system (Baker et al., 2013) behaviors that are associated with poor learning outcomes. Similarly, log-file data recorded from an online learning platform was used to examine learner participation by utilizing the characteristics of each post to predict learners' level of engagement (Shukor et al., 2014). Results from this study found that the most effective predictors of engagement were metrics about posts where learners shared information or posted high-level messages (i.e., elaborative text).

Other studies have used log-file data to identify learner profiles of cognitive engagement during learning. Kew and Tasir (2021) analyzed log files to identify behaviors of low and high cognitive engagement displayed by learners. This study defined engagement by the quality of posts on an educational forum where each learner's cognitive engagement level was determined by comparing the ratio of low-level cognitive contributions (e.g., providing an answer to a post without explanation) to the ratio of high-level cognitive contributions to the e-learning forum (e.g., providing an explanation). Learners were identified as having either high, high-low, or low cognitive engagement depending on the relationship between the proportion of high to low cognitive engagement displayed in their posts. Findings from log-file data found that most learners were categorized as having low cognitive engagement on online forums, providing insight as to how to encourage cognitive engagement through e-learning platforms. Similarly, Li et al. (2020) identified profiles of learners based on their log-file data as they learned with BioWorld, a simulation-based training environment. Findings from latent profile analysis revealed several different types of cognitive engagement including recipience, resource management, and task-focusing (Corno & Mandinach, 1983). Additionally, results found that learners who were categorized as either resource management or task-focus cognitive engagement had greater diagnostic efficacy than learners with recipience cognitive engagement.

Log files, as demonstrated in the studies above, are revealed as important indicators and measures of engagement. Using log files has several advantages for researchers as log files: (1) are unobtrusive process-based data that can be collected online within traditional (e.g., classroom) or nontraditional (e.g., virtual) learning environments; (2) gather rich temporal data that can be contextualized, allowing for sophisticated analytical techniques to be used for examining individual learners' time-series data; and (3) should the task align with engagement theories, log files can serve as accurate identifiers of engagement during learning. However, there still exist limitations in using solely log-file data to capture, measure, analyze, and interpret cognitive engagement. Log files require interaction or physically expressed behaviors to capture engagement and as such is better suited to capture behavioral engagement rather than cognitive or emotional engagement. Historically, log files have been used to make inferences about cognitive engagement, but these inferences must be theoretically justified (Azevedo, 2015). Additionally, it is currently unknown in the literature at what time log files may be used to best capture engagement or at what sampling rate engagement indicators are unreliable or unable to be aligned with other data channels.

4.2 Eye Tracking and Gaze Patterns

Eye-tracking data refers to the experimental method of recording learners' gaze behaviors, including fixation points, saccades, regressions, and dwell times, as they engage in a task (Carter & Luke, 2020). Using eye-tracking data, researchers can identify where a learner looks, for how long a learner gazes at an area of interest (AOI; i.e., a region of an object to contextualize where a learner is looking), how often they move from one AOI to another, and the sequences a learner gazes at a battery of AOIs. Eye tracking has been used across multiple studies to capture learner engagement during reading tasks (Miller, 2015), learning with virtual environments (Bixler & D'Mello, 2016; Wang et al., 2020), designing cueing animations (Boucheix et al., 2013), etc.

Miller (2015) equated eye-tracking data, more specifically dwell times (i.e., aggregation of fixation durations) to increased thinking and attention on specified AOIs. For example, a learner who has a greater dwell time on one object would be assumed to have been thinking about the object more than an object where dwell time was lower. However, there is a large assumption being made – a learner is not engaged with material if they are not looking at the material and they are engaged if they are looking at the material. As such, studies have attempted to examine mind-wandering patterns in relation to learning outcomes using eye-tracking data. Mind wandering, also known as zoning out, is an unintentional attentional shift toward non-task-related thoughts (Killingsworth & Gilbert, 2010). Bixler and D'Mello (2016) used eye tracking to detect when a learner demonstrated mind-wandering behaviors during a reading task. During this reading task, participants were asked both during a passage and at the end of a page to report occurrences of mind wandering while calibrated to an eye tracker. Using machine learning techniques, mind wandering was detected with 72% accuracy. This study highlights the importance of contextualizing psychophysiological data to ensure appropriate interpretations of the data are being made.

Engagement with instructional materials was similarly detected using eye tracking by D'Mello et al. (2012) and integrated with an intelligent tutoring system, Gaze Tutor, to provide learners scaffolding during a task. Individual learners' gaze battens were used to identify if the learner was disengaged to then prompt the learner via dialog to reengage the learner. Findings of this study reported increased learner attention through the use of gaze-sensitive dialogues. Similarly, Bidwell and Fuchs (2011) identified individual learners as either engaged or disengaged using eye tracking where learners were classified into one of three states: engaged, attentive, or resistive. However, when compared to expert human coders' classification of student engagement, hidden Markov models were only 40% accurate in classifying learners, perhaps highlighting the role and impacts of subjectivity of subjects and observers in some research.

These studies highlight both the strengths and limitations of using eye tracking to identify and measure engagement. The method of collecting eye-tracking data can be expensive and intrusive to the learner due to the calibration and equipment setup. Additionally, the collection can be complex as it may be affected by the individuals' physical actions such as sweating or moving (Henrie et al., 2015). Although collecting this data can be difficult, eye-tracking data can be collected at multiple levels of granularity, from milliseconds to hours of aggregation and timespans. In addition, eye tracking can measure temporal sequences of actions through saccades, attention allocations via fixation durations and dwell times, and cognitive effort through pupil dilation, providing researchers with richly quantified dataset contextualized to the learning task (e.g., see Dever et al., 2020; Taub & Azevedo, 2019; Wiedbusch & Azevedo, 2020).

4.3 Audio/Video (Think and Emote-Alouds, Observations, and Interviews)

Audio and video serve as methods for collecting think-alouds (i.e., concurrent verbalizations) of learners' thoughts as they complete a task (Ericsson & Simon, 1984), emote-alouds where verbalizations consist of emotions, observations of learners' actions during a task, and interview data for post-task qualitative analysis (D'Mello et al., 2006). Data from audio and video can provide insight as to how learners demonstrate engagement with material during a task and provide critical contextual cues needed to make accurate inferences about engagement. For example, Tausczik and Pennebaker (2010) argue that word count calculated through think-aloud audio data can identify a learner who is dominating a conversation with a peer, teacher, or tutor as well as the level of engagement that is demonstrated by a learner (e.g., high, low).

Past studies have examined think-alouds across contexts and domains to identify engagement. For example, one study used a combination of interviews and video as learners completed math lessons to measure cognitive engagement (Helme & Clarke, 2001). Findings revealed that cognitive engagement was accurately identified through both linguistic and behavioral data from audio and video data respectively. Linguistic indicators of cognitive engagement included verbalization of thinking, seeking information, justifying an argument, etc., where behavioral indicators were primarily identified through gestures. Another study used audio data to identify linguistic matching during a negotiation between multiple parties (Ireland & Henderson, 2014). Within this study, lower task engagement levels were associated with an increase in language use and style matching (i.e., percentages of words in various linguistic categories) but were indicators of higher social engagement. More specifically, the mimicry of verbal and non-verbal communication showed an increased attention to social cues but had a negative relationship with task engagement as pairs were more likely to hit conflict spirals and impasses. A study by Ramachandran et al. (2018) also examined social engagement via audio data where the word count and the number of prompts were recorded as a conversation took place between a learner and a robot tutor. Specifically in this study, learners were required to think aloud and while doing so, a robot tutor would prompt the learner to consistently think aloud Using a robot-mediated think-aloud showed improvements in students' engagement and compliance with the think-aloud protocol compared to using just the robot without think-aloud prompting, tablet-driven think-aloud prompting, or neither the robot or think-aloud prompting, indicating the potential value of using social robots in education for (meta)cognitive engagement.

While audio and video data can serve as a non-intrusive method of rich data collection to measure cognitive, behavioral, task, and social engagement, there exist several limitations that are specific across different data collection methods (Azevedo et al., 2017). Observational methods can be expensive and require trained and paid professionals. For example, the BROMP coding technique (Baker et al., in press) is a momentary time sampling method in which trained certified observers record student's behavior and affect in a pre-determined order using an app that can then automatically apply various coding schemes. In addition, think- and emote-alouds require learners to be able to accurately and consistently verbalize their thoughts, emotions, and cognitive processes which may slow performance as they try to complete a task sometimes complex in nature (e.g., problem solving, learning about a difficult concept). Further, despite the density of utterances, these studies using these methods tend to have smaller sample sizes, making it difficult to generalize to other studies. However, audio and video data can focus on the activity level, provide qualitative aspects of engagement (e.g., emotional engagement), and contextualize other data type measures of engagement such as eye-tracking data. They also focus on the veracity of the data such that while the number of subjects may be more limited, the depth of the data collected is rich and offers a valuable corpus that can then be inspected from multiple vantages.

4.4 Electrodermal Activity and Heart Rate Variability

Electrodermal activity (EDA) data manifests from changes in learners' topical electrical conductance, quantifying sweat gland activity to identify stimuli such as cognitive engagement (Posada-Quintero & Chon, 2019; Terriault et al., 2021). Heart rate variability (HRV) measures the fluctuation of the duration between heartbeats to identify the temporal relationship and changes in sympathetic and parasympathetic effects on heart rate (Appelhans & Luecken, 2006). Both physiological data channels aim to mitigate the limitations of traditional techniques such as survey-based measures that can be time-consuming and cognitively demanding for the learner (Gao et al., 2020). Because of this, studies have attempted to understand how non-invasive EDA and HRV data collection methods can be used to capture and measure learner engagement.

A study by Gao et al. (2020) explored how learners' cognitive, behavioral, and emotional engagement could be captured by EDA metrics and which of those metrics are the most useful in predicting learner engagement as well as differentiating between the three types of engagement. Results from this study found that cognitive, behavioral, and emotional engagement level during class instruction can be detected with 79% accuracy across 12 EDA metrics in addition to other physiological metrics (i.e., photoplethysmography, accelerometer). In examining the relationship between EDA peak frequency and the three types of engagement, Lee et al. (2019) found that a greater number of peaks indicating increased arousal was related to greater cognitive and behavioral engagement. However, in relating EDA peak frequency to emotional engagement, the study did not find significant associations possibly due to those activating emotions, either positive or negative, that can have both positive and negative relationships with emotional engagement (Lee et al., 2019). In contrast, Di Lascio et al. (2018) found that when measuring emotional engagement during class, increased levels of arousals were related with greater levels of emotional engagement.

As seen in the slight variation in findings across studies, collecting EDA and HRV data can be challenging due to the limitations presented for data collection, analysis, and implications. Specifically, both data channels can be intrusive due to the instrumentation of learners that must occur. While some instruments, like a smart watch, can unobtrusively collect this information, more sophisticated and expensive instruments allow for greater accuracy (e.g., greater sampling rate; Henrie et al., 2015). Additionally, many precautions and considerations must be taken in both the environmental conditions (e.g., temperature) and participants' individual physiological and lifestyle differences (e.g., weight, caffeine and medication consumption, etc.; Terriault et al., 2021). Interpreting arousal via EDA and HRV data can be difficult without the use of additional data channels such as within the study by Di Lascio et al. (2018) who compared arousal data against self-report measures to triangulate the validity of arousal measures and implications. While these data channels can be used to accurately predict levels of cognitive and behavioral engagement in learners during a task or lecture, emotional engagement has yet to be concretely identified through these techniques. However, EDA and HRV methods collect rich, fine-grained data that allow researchers to create individualized models of engagement.

4.5 Self-Reports and Experience Sampling

Self-reports and experience sampling have been long-standing measures of cognitive, behavioral, and emotional engagement due to the ease of administration and the ability to understand learners' reflections on their engagement. To obtain these data, learners are asked to report experiences and understanding of their own degree of engagement during a learning task either prior to (e.g., "Before a quiz or exam, I plan out how I will study the material"; Miller et al., 1996) or after (e.g., "To what extent did you engage with the reading material?") their task. Several studies have not only developed scales for engagement (e.g., Appleton et al., 2006; Vongkulluksn et al., 2022) and examined these scales for reliability and accuracy (Fredricks & McColskey, 2012), but have also examined and determined learners' level of engagement using self-reports and experience sampling. A study by Salmela-Aro et al. (2016) used the experience sampling method of short questionnaires throughout a science class across 443 high school students. From this sample and using latent profile analysis, this study found four profiles of learners – engaged, engaged-exhausted, moderately burned out, and burned out. Through this assessment and methodology of data collection, this study was able to examine both positive and negative aspects of engagement. Xie et al. (2019) also used experience sampling to measure cognitive, behavioral, and emotional engagement across several self-report measures of engagement. Findings from this study established event-based sampling as a more accurate way that cognitive, behavioral, and emotional engagement can be captured by self-reports. Finally, experience sampling type of self-reports allowed researchers to have a deeper exploration of how engagement relates to learner behaviors.

Using self-reports and experience sampling methodologies to capture, collect, analyze, and interpret cognitive, behavioral, and emotional engagement demonstrated by learners has several strengths The method is easy and cheap to administer to learners and provides a representation of learner reflection and perception of engagement during a task (Appleton et al., 2008). Additionally, these methods can be used to compare across scales and, as Appleton et al. (2006) argue, can be the most valid measure of both cognitive and emotional engagement as both constructs rely on learners' self-perception. However, a review of cognitive engagement self-report measures by Greene (2015) showed that researchers have begun to over rely on the information provided by these measures, without regard to the several limitations these metrics pose. For example, self-report measures of engagement have not fully developed the definition and multidimensional conceptualization of cognitive, behavioral, and emotional engagement, leading to a divided field regarding the indicators of engagement during learning (Fredricks & McColskey, 2012; Li et al.,

2020). Additionally, the assumption is made due to self-reports that engagement is stable across time, can be aggregated and misaligned with real-time behaviors demonstrated by learners, and can be measured outside of the immediate learning task (Greene, 2015; Greene & Azevedo, 2010; Schunk & Greene, 2017). One-way studies have attempted to rectify this limitation is through the prompting of self-reports throughout the learning task. However, this prompting can be disruptive to the learner as well as cognitively demanding during a learning task (Penttinen et al., 2013). As such, several pieces of literature have indicated self-reports (generally) as poor indicators of the construct that was intended to be measured (Perry, 2002; Perry & Winne, 2006; Schunk & Greene, 2017; Veenman & van Cleef, 2019; Winne et al., 2002).

4.6 Facial Expressions

Facial expressions have primarily been used to identify learners' internalized and temporal emotions as they complete learning tasks. To do so, video clips of learners are captured and enumerated using several different algorithms which identify different states of emotions including happiness, anger, joy, frustration, boredom, etc. One example is the Facial Action Coding System (FACS; Ekman & Friesen, 1978) which maps action units, or specific landmarks, onto the learner's face to monitor and quantify which facial structures move, when they move, and in conjunction with other action units. From this, emotion scores are derived which indicates the probability of an emotion being present.

Several studies have used machine learning techniques on learners' facial expressions to identify at what point of time and the duration a learner demonstrates engagement on a learning task (Grafsgaard et al., 2013; Taub et al., 2020). A study by Whitehill et al. (2014) examined methods to automatically detect instances of engagement using learners' facial expressions in comparison to human observers judging emotions displayed by learners in 10-second video clips. Findings of this study established machine learning as a valid technique for reliably detecting when a learner displays high or low engagement. Similarly, Li et al. (2021) employed supervised machine learning algorithms to identify how learners demonstrated cognitive engagement using facial behaviors as they deployed clinical reasoning in an intelligent tutoring system. Results found that three categories of facial behaviors (i.e., head pose, eye gaze, and facial action units) can accurately predict learners' level of cognitive engagement. Moreover, there were no significant differences in the overall level of cognitive engagement between high and low performers. However, learners in this study who were classified as high-performance demonstrated greater cognitive engagement as they completed deep learning behaviors.

Prior literature has shown that engagement can be detected and predicted using learners' facial expressions. However, using facial expressions to identify engagement assumes that all emotions are depicted by facial expressions. More specifically, we ask the question: do all emotions need to be expressed facially to exist?

From this, there is a limitation to determining the level and type of engagement as emotions could potentially be completely internal without outward indicators of their presence. Using facial expressions as a measurement of emotion assumes facial expressions are universal while ignoring potentially important social constructs (e.g., culture, positions of power, dynamics of relationships, etc.) and context. For example, a smile might not always indicate happiness if following bad news. In this case, the smile may be interpreted as an emotion-regulatory strategy or dismissive strategy to negative emotions. It may also indicate mind-wandering or disengagement if the smile is not related to any event or trigger from the environment or learning task.

Despite these limitations, facial expressions have been shown to be accurate and reliable indicators of engagement according to past studies (e.g., Grafsgaard et al., 2013; Li et al., 2021; Taub et al., 2020; Whitehill et al., 2014). Facial expressions, in addition to reliable detection, are non-invasive and able to be automatically coded in real time without the utilization of human resources.

4.7 EEG

Electroencephalogram (EEG) is a physiological measure of summed postsynaptic potentials as neurons fire that provide temporal information about dynamical changes in voltage as measured via electrodes attached to the scalp and a reference electrode (Gevins & Smith, 2008). From these electrode voltage signals, there are several frequency bands that have been used to measure cognitive states and processes such as vigilance decline (Haubert et al., 2018), information processing (Klimesch, 2012), and mental effort (Lin & Kao, 2018).

Pope et al. (1995) developed an engagement index based on a ratio between the amplitudes of beta, alpha, and theta frequency bands that was found more sensitive to changes in cognitive workload demands than other indices. This index has since been reconfirmed as a sensitive measure to cognitive engagement during various cognitive lab-based tasks (Freeman et al., 1999; Nuamah & Seong, 2018), and has been used to study engagement in children during reading (Huang et al., 2014), employees in workplaces (Hassib et al., 2017a), and university students during lectures (Hassib et al., 2017b; Kruger et al., 2014).

Studies using EEG are less intrusive than other brain-scanning methods and have been conducted within lectures using headsets (e.g., Kruger et al., 2014). These studies provide temporally rich and fine-grained data that are prime for measuring cognitive engagement but are computationally and resource intensive. Additional types of non-intrusive brain-scanning methods (e.g., fNIRS) have been used to detect features of engagement (e.g., Verdiere et al., 2018) using similar operationalizations of cognitive engagement which may prove to be a synergistic measurement tool to EEG in future work examining other dimensions of engagement.

4.8 Convergence Approaches

Above we reviewed several unimodal data channels that have been used to study engagement, with many of those studies using a second data channel to validate newer measures (e.g., Bixler & D'Mello, 2016; Gao et al., 2020; Taub et al., 2020). As important, the use of multiple data sources also (1) provides complementary information when used in conjunction with one another (Azevedo & Gašević, 2019; Azevedo et al., 2017; Sinatra et al., 2015), (2) provides contextual information for interpretation (e.g., Järvelä et al., 2008), and (3) can be used to develop more holistic models by identifying interrelations among related variables (Papamitsiou et al., 2020).

For example, Dubovi (2022) measured the emotional and cognitive engagement of learners using a VR simulation of a hospital room using facial expressions, self-reports, eye-tracking, and EDA. It is important to note, however, that in this study these channels were all analyzed independently. That is, these two metrics were not combined into a single "emotional engagement" metric but rather the authors report that facial expressions were used to examine dynamical fluctuations in fast-changing emotions while the self-report measured the intensity of emotions at set times throughout the task. This study shows one way to use multiple data channels to complement one another during interpretation.

Attempts have also been made to evaluate engagement using multimodal data that are used in conjunction with one another to help provide additional context. For example, Sharma et al. (2019) used both head position and facial expression to classify learners' engagement level. Their system begins by detecting the face and head position to determine the learners' attentional state (i.e., distracted or focused), and if the learner is focused, the dominant facial emotion is measured, and an engagement level is calculated based on the dominant emotion probability and a corresponding emotion weight. This value corresponds to a classification of "very engaged", "nominally engaged", and "not engaged". This approach underscores how one data channel can be used to provide contextual information for another data channel measurements to occur.

Finally, other studies have attempted to fuse multimodal data to develop more holistic measures of engagement. Papamitsiou et al. (2020) were able to use log files, eye-tracking, EEG, EDA, and self-report data in a fuzzy set qualitative comparative analysis approach (using 80%, 50%, and 20% thresholds for degree of membership) to create a multidimensional pattern of engagement. Their approach identified 9 configurations of factors to help explain performance and engagement on a learning activity. Their findings showcase how multimodal data fusion suggests more than one pattern of engagement that facilitates higher learning outcomes. That is, because engagement is a multidimensional construct, there are likely multiple avenues by which engagement impacts learning and that models of the measurements of engagement should reflect this.

The promising shift toward multimodal approaches for measuring engagement is largely driven by new analytical techniques such as those described above. However, there are limitations around using a multimodal approach. We discuss these limitations in further detail later in this chapter (see Sect. VI. Limitations and Future Directions) but highlight that these studies are not easy to collect for a large number of participants. As such, many of these studies have relatively smaller sample sizes that should be considered when discussing generalizability across learning context. As new multimodal analytical techniques emerge, we can expect to see a large increase in these types of studies which will help address this concern. It is imperative that this work, however, be theoretically driven to avoid data hacking or phishing expeditions. Grounded within this model.

5 Theoretically Grounded Approach for Measuring Engagement with Multimodal Data

As we have previously shown, there have been many approaches to studying engagement. However, the data that currently has been used to measure engagement have varying conceptualizations and degrees of utilizing the multifaceted definition of engagement we outlined previously. This means that a direct comparison of these methods' effectiveness and efficacy is not only difficult but also ill-advised. Instead, we suggest future research needs to be explicit in what components and facets of engagement are of interest to help advise which channels would be deemed most appropriate. That is, there is likely not a single channel that will provide a single metric for best quantifying engagement. Rather, each channel has its strengths and limitations for each component and facet of engagement that must be considered. These considerations can include which phase of SRL is of interest (i.e., forethought, performance, or self-reflection), the facet of engagement that is being measured or inferred (e.g., cognitive versus behavioral versus emotional), the temporal granularity (e.g., changes in engagement moment to moment versus sub-goal to sub-goal versus day to day), the context as constrained by the environment, and combination of converging data channels (see Azevedo et al., 2017, 2019). Using multimodal data to measure psychological constructs is not a novel approach, as we have highlighted in several studies above. Often, multiple channels are used to validate another (e.g., using self-report data to validate EDA fluctuations of arousal indicating higher engagement). However, multimodal data can also be used in conjunction to provide complementary information for measuring engagement as well (D'Mello et al., 2017). Importantly though, how best to integrate and utilize multiple channels is not yet fully understood and requires additional empirical work that is grounded in a unifying model of engagement.

Below, we provide examples of how these methods could be used specifically to interpret not only each facet of engagement, but also their individual components as situated in our extended integrative model of SRL engagement (see Table 10.1). This table highlights how no one channel can capture all components of all facets of engagement. For example, facial expressions might provide fine-grained data on

Table 10.1 Data channels and examples for measuring engagement within each phase of SRL

Scenario: A learner is using an intelligent tutoring system to learn everything they can about Topic X given a specific time constraint (e.g., see Azevedo et al., 2022).

Data Channel	SRL phase and components	SRL facets	Examples	Exemplar citations (utilizing the data channel)
Eye-tracking	Forethought	Initiating cognitive engagement (strategic planning, effort planning)	Gaze patterns are used to determine how the learner scans the navigation toolbar which lists the various tools available to use (e.g., a notepad, quiz, video tutorials).	Antonietti et al. (2015), Boucheix et al. (2013), D'Mello et al. (2017), Duchowski (2007), Miller (2015) and Van Gog and Jarodzka (2013)
	Performance	Enacting behavioral engagement	Dwell times are used to determine how long the learner spends looking at the video tutorial.	
	Self-reflection	Evaluating B/E/C engagement	Gaze on relevant performance or learning metrics following the learning session provided by the system are used to determine if the learner is reflecting on their previous performance phase.	
Log files/click streams	Forethought	Initiating cognitive engagement (strategic planning, effort planning)	Event markers that are quick in succession are used to determine when and which tools the learner opens up to scan to determine what is available to use as they orient to the learning environment.	Azevedo et al. (2015), Bernacki et al. (2012), Gobert et al. (2015), Henrie et al. (2015), Järvelä et al. (2008), Kew and Tasir (2021), Li et al. (2020, 2021) and Shukor et al. (2014)
	Performance	Maintaining cognitive engagement	Dwell times are used to determine how long the learner spends looking at the video tutorial	
		Enacting behavioral engagement	Keylogging captures the notes a learner's take as they watch a video tutorial.	
	Self-reflection	Evaluating B/E/C engagement	Event markers that are quick in succession to previously visited pages or personal notes are used to evaluate the past performance phase.	
		Adjusting B/E/C engagement	State transitions inferred via modeling approaches (e.g., Markov chains) or event analysis may indicate points of reflection requiring intervention from the learner.	

(continued)

Table 10.1 (continued)

Scenario: A learner is using an intelligent tutoring system to learn everything they can about Topic X given a specific time constraint (e.g., see Azevedo et al., 2022).

Data Channel	SRL phase and components	SRL facets	Examples	Exemplar citations (utilizing the data channel)
	Manifestations	Absorption, dedication, vigor, performance	The change in frequency or rate of actions a learner takes may indicate changes in the level of effort that are being put forth. Additionally, log files can be used to see specific steps taken in assessment-based questions to quantify performance.	
Facial expressions	Antecedents/facilitators	Mood	Evidence scores for the level of joy or frustration are captured prior to the learning session to determine how a learner feels prior to interacting with the environment.	Grafsgaard et al. (2013), Li et al. (2021), Taub et al. (2020), Whitehill et al. (2014)
	Task-related emotions	Interest, enthusiasm, curiosity, frustration, distress, anxiety	Evidence scores for frustration and confusion are captured throughout the learning session to track how a learner feels at various stages of learning.	
Emote Alouds	Antecedents/ facilitators	Mood	The frequency of verbalizations where the learner indicates their mood ("I feel good today!") is calculated	Baker et al. (2013) and Craig et al. (2008)
	Task-related emotions	Basic academic-related emotions (frustration, confusion, anger, joy)	The frequency of verbalizations in which a learner indicates an emotion is calculated. ("I am frustrated I don't understand topic X!")	

Scenario: A learner is using an intelligent tutoring system to learn everything they can about Topic X given a specific time constraint (e.g., see Azevedo et al., 2022).

Data Channel	SRL phase and components	SRL facets	Examples	Exemplar citations (utilizing the data channel)
Think Alouds	Antecedents/ facilitators	Motivation, belief	The frequency of verbalizations indicating how motivated a learner is is calculated. ("I am really excited to learn about topic X, it has always been a curiosity of mine.")	Azevedo et al. (2022), Appleton et al. (2006), Dent and Koendka (2016), Duffy and Azevedo (2015), Fredricks and McColskey (2012), Greene et al. (2015), Helme and Clarke (2001) and van der Graaf et al. (2022)
	Forethought	Task analysis (goal setting)	The frequency of verbalizations of explicit goals is calculated. ("I want to learn specifically about subject Y within topic X.")	
		Initiating cognitive engagement (strategic planning, effort planning)	The frequency of verbalizations of plans is calculated. ("first, I'll see what videos are available since I'm new at this.")	
	Performance	Maintaining cognitive engagement	The duration of pauses in between verbalizations of effort is calculated to compare over time.	
		Monitoring cognitive engagement	The frequency of verbalizations of effort is calculated. ("this practice problem is requiring me to try really hard.")	
	Self-reflection	Evaluating B/E/C engagement	The frequency of verbalizations of reflection is calculated. ("I think the note-taking tool was not very helpful for this task.")	
		Adjusting B/E/C engagement	The frequency of verbalizations of changes to levels of interest is calculated. ("I'm not as interested in topic X as I used to be.")	

(continued)

Table 10.1 (continued)

Scenario: A learner is using an intelligent tutoring system to learn everything they can about Topic X given a specific time constraint (e.g., see Azevedo et al., 2022).				
Data Channel	SRL phase and components	SRL facets	Examples	Exemplar citations (utilizing the data channel)
Self-reports	Antecedents/facilitators	Motivation, belief, mood	Goal orientations are measured prior to learning.	Fredricks and McColskey (2012), Greene (2015), Renninger and Bachrach (2015), Vongkulluksn et al. (2022) and Wolters (2004)
	Self-reflection	Evaluating B/E/C engagement	Retrospective confidence judgments are used to measure a learner's behavioral engagement use.	
	Task-related emotions	Interest, enthusiasm, curiosity, frustration, distress, anxiety	Scales for curiosity are administered after a new sub-goal is indicated and used to determine changes in task-related emotions	
	Manifestations	Absorption, dedication, vigor, performance	Scores on a quiz after sub-goal completion are administered throughout the learning session to examine fluctuations over time	
Eda	Performance	Maintaining cognitive engagement	The frequency of high arousal peaks during video lectures is compared to determine which videos were most emotionally engaging.	Dubovi (2022), Gao et al. (2020), Di Lascio et al. (2018), Lee et al. (2019), McNeal et al. (2020) and Terriault et al. (2021)
	Manifestation	Absorption, dedication, vigor	Increases in percent change of skin conductance are used to infer changes in vigor.	
Heart rate	Antecedents/facilitators	Mood	Heart rate variability is used to infer emotional states that are captured prior to a learning session to serve as a baseline for a learner's current mood.	Gao et al. (2020) and Wang and Cesar (2015)
	Performance	Maintaining cognitive engagement	The fluctuations in heart rate variability are used to determine changes in arousal levels during various strategy uses.	
	Task-related emotions	Interest, enthusiasm, curiosity, frustration, distress, anxiety	Heart rate variability is used to infer emotional states via arousal during learning.	

one's expressed task-related engagement emotions as they fluctuate with their interactions between all SRL phases, but they would provide little to no information about behavioral engagement (albeit context may be inferred from the facial expressions to provide some interpretation of behaviors such as why someone might be engaging in a particular strategy).

This table highlights a couple of interesting challenges when working with multimodal data that go beyond what has already been published (see Azevedo et al., 2017, 2019, 2022; Järvelä & Bannert, 2021; Molenaar et al., 2022). First, many of these data are starved of qualitative information that can be derived from another channel. For example, this table describes an example of how log-file data could be used to evaluate engagement during the self-reflection phase. We suggest that event markers that are quick in succession to previously visited pages or work can be used to indicate reflection. Additionally, this table highlights how events are still being examined primarily independently of one another instead of thinking of actions or events that are more communal. That is, much in the same way we can use collections of facial landmarks to detect faces and facial expressions, log-file event markers could be used to create constellations indicating various types of engagement. However, unaccompanied by think-alouds or self-reports, what the learner was consciously reflecting on may not be differentiated from unconscious reflection of what was being reflected upon. That is, the learner could be consciously reviewing the length of their notes or determining if they had seen all of the material by flipping through the informational pages of an environment, but unconsciously evaluating how much effort taking those notes or reading all of that material took and whether or not they felt it was effective to their current judgment of their learning. Log files would not be able to make this distinction alone, but rather must be inferred (perhaps based on the type of content being reviewed or the order the content is reviewed). The addition of think-aloud data might provide more context as to the *why* of the reflection.

This table also highlights where the various channels benefit from the contextualization of other channels. That is, the same metrics might be recorded and not be able to provide interpretable delineation between the SRL phases without other data channels. For example, our table suggests that heart rate variability can be used to determine task-related emotions and the maintenance of cognitive engagement. However, it is important to note that this requires a level of inference-making as the metric is reporting on arousal. Additional context is needed to understand if fluctuations in the heart rate variability are driven by changes in effort (indicating performance-phase cognitive engagement maintenance) or changes in task-related emotions (e.g., anxiety or distress). By introducing additional data, such as speech or self-reports, important distinctions can be made. That is, the addition of self-report data might provide more context on the *when* or *what SRL phase* of physiological data.

As we consider multimodal data channels for inferring engagement levels of learners, we must also consider not only their temporal granularity in relation to one another (i.e., eye-tracking versus self-report measures) but also within each channel (Azevedo & Gašević, 2019), and potentially in the "fused" channels. For example,

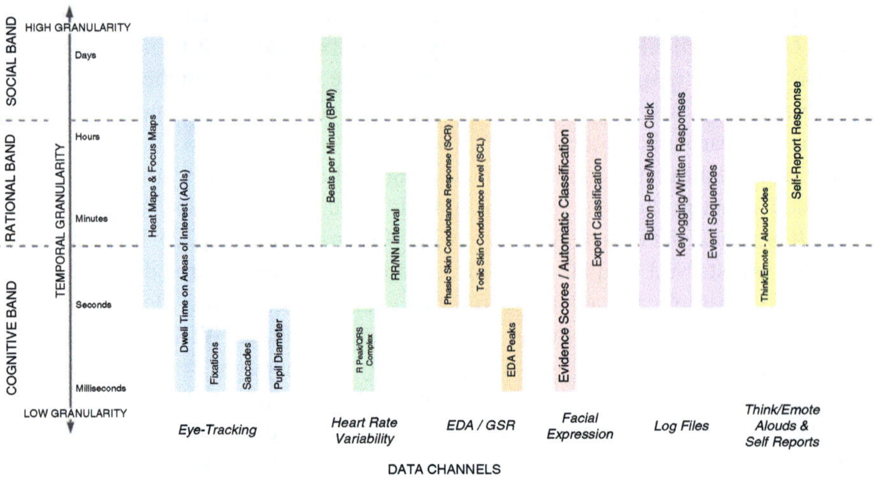

Fig. 10.2 Temporal banding of multimodal data channels for measuring engagement

within eye-tracking, we must consider the inferential implications of using more fine-grained data (e.g., fixation durations) compared to more aggregated forms of data (e.g., heatmaps). This granularity is further explored in Fig. 10.2, which is a non-exhaustive set of metric examples that can be used to make inferences about engagement. These data can be used to explore when, how long, how often, and the shape or topography of occurrences transpire during the various fluctuations of engagement within the SRL phases outlined in our model.

Within each data channel (horizontal axis), we provide examples of metrics (colored bands) that can be collected (e.g., timestamps of button presses from log files) or generated (e.g., Think/Emote-aloud codes). These have been situated along the vertical-hand's scale of more fine-grained and typically raw data up to more aggregated data. For example, fixations and saccades are collected within the millisecond range, but can be accumulated and aggregated across minutes or hours to determine dwell times. This figure outlines the relative temporal banding of the exemplar metrics that can be used to make inferences about learners' cognitive, behavioral, and emotional engagement during SRL. These inferences are best captured with multimodal data using a variety of data channels (e.g., eye-tracking) with multiple metrics (e.g., fixations, dwell time), and variables (e.g., fixation frequency, dwell time duration) that can be extracted to evaluate engagement while learning. As research continues to develop novel and innovative approaches to measuring various psychological constructs using both online trace data and offline sources, the metrics available for use are a growing list (see Darvishi et al., 2021). Those outlined above just scratch the surface at what has been previously examined, but we acknowledge that many more metrics exist that could fit well into our model. Additionally, within each of those metrics, there are many variables that can be extracted. Almost all metrics can be analyzed using frequency (i.e., how often something occurs), duration (i.e., how long it took to occur), and their timing within the learning context/timeline (i.e., when something occurs).

The dimension of time scale of grain size has further been split (horizontal dashed lines on across the vertical axis) separating the cognitive, rational, and social bands on the left vertical) according to Newell's (1994) levels of explanation corresponding to the time scale of human actions, to highlight which data are most appropriate when making inferences about cognitive (unit tasks, operations, and deliberate acts), rational (task level activity), or social (e.g., course-length engagement) activities. This work was in large part to help the development of cognitive architectures, and as such these bands represent qualitative shifts about the type of processing assumed to occur within them and the manner researchers talk about their internal levels from a systems-level perspective (West & MacDougall, 2014). Briefly, the cognitive band represents symbolic information processing, the rational band represents the level knowledge becomes abstract to create a (imperfect) knowledge level system, and the social band refers to distributed multi-agent processing.

However, it is important to note that the delineations between the bands are not hard boundaries but rather gradual guidelines (as indicated by being dashed and not solid). Furthermore, according to Anderson's (2002) "Decomposition Thesis", there is much evidence that suggests human action occurring at grander time scales are composed of smaller actions at shorter time scales. That is, most of what occurs in the social band involves a great degree of rational and cognitive processing. This figure highlights how it is important in the work around engagement that we must be clear about the temporality of interest when discussing the quality and quantity of engagement. For example, are we concerned with fine-grained attentional shifts within a single task or the overall level of interest and emotional investment of a semester-long course consisting of multiple lectures each with multiple tasks?

6 Limitations and Future Directions

This chapter provides groundwork for future engagement research by drawing new connections between associated constructs and measures. Specifically, the narrative offers a multifaceted theoretical conceptualization of engagement and associated data channels. Considering broader implications and future directions for engagement-sensitive AIEd systems, we see an additional strength of our model (Azevedo & Wiedbusch, 2023), in that the interaction between the individual learner and the environment is one that allows for feedback loops. These have the potential to become externalized and therefore are amendable to IMMSE analysis. For example, we can imagine a system that detects waning cognitive and behavioral engagement during strategy use within the performance-phase based upon eye-tracking and log file data. Upon this detection, the system may then choose only at that moment to interrupt the learner to probe about their current emotional engagement levels and offer suggestions how best to increase levels of interest or curiosity. In this way, the system is directly adapting to the user. However, we must remember these systems should also be used to scaffold learners, so some intrusions need not be only measurement-related in nature. These intrusions can be intervention-driven and serve as additional data-rich sources for future measurement without this being their main intent. Additionally, as the user continues to interact with the system, we can imagine that it begins to track which interventions prove to be the most successful in increasing engagement. These interventions are then made more readily available for the user within the system while also suppressing those interventions that have been shown to decrease the individual's level of engagement. In this way, the learner is directly adapting the task and environment affordances and constraints to improve their learning experience. Due to learner individual differences, these changes could be made in such a way that no two learners' environments are the same.

Future work can also test and elaborate on specific connections forwarded in the IMMSE, including how prior knowledge, task constraints, and goal setting influence engagement facets. For example, the present model (Fig. 10.1) highlights connections between task analysis (esp. goal setting) and the initiation and maintenance of cognitive engagement, which can be measured via gaze fixations and EDA, among others. Connections such as these can be empirically examined, not only to test the model, but to forward appropriate SRL interventions and measures of cognitive engagement. Additional work can also more thoroughly address the role of agentic engagement within the context of self-regulated learning, which should be expanded upon in future model iterations.

This work also recognizes the general advantages and disadvantages of multimodal approaches to assessing engagement, as well as the need for ongoing research in this broad area. A conclusion one can draw from the IMMSE model, which highlights a tension between collecting as much as possible, and knowing which channels are most helpful to a particular context and analysis, is that the model will encourage more research that contrasts the relative utility of different measurement channels and metrics – solo and combined – in studying particular constructs in particular contexts (e.g., Amon et al., 2019, 2022). For example, it is increasingly popular in the realm of multimodal measurement as a sensory-suite approach to research, where all available measurement channels are utilized during research studies within a given lab, even if a particular measurement channel is not central to the motivating research questions. Research is conducted in this fashion for good reason: Research is expensive and time-consuming, and – for those fortunate enough to afford such setups – elaborate sensory suites provide more "bang for the buck." By capturing as much information during a study as possible, researchers can push creative research questions to the forefront and harvest data for years to come. Certainly, the sensory-suite approach is a good investment in many cases, but it has some caveats. Researchers may put the cart before the horse in terms of research outputs, feeling inherent pressure to forward all data channels as useful in a given context (e.g., in terms of predictive value) or present a multimodal approach as better than a unimodal approach without appropriate testing. For instance, a researcher may hesitate to disseminate findings that EDA has negligible predictive value compared to eye tracking, if the researcher has intentions to continue submitting papers centered on EDA results and may instead present only data that supports the multimodal approach. Additionally, there are still many methodological questions around the generalizability of multimodal approaches and their data sampling to subject rates. In what contexts is having 100,000 samples of one individual better or worse than 1 sample from 100,000 subjects? Where should researchers attempt to strike the balance between generalizability and data veracity? In the long term it is pragmatic and prudent the field begins to hone in on specific best practices in multimodal (or unimodal) engagement measurement.

Lastly, the present work has several limitations, including depictions of the IMMSE ongoing task dynamics. Whereas delineated boxes may suggest discrete stages, they likely overlap. We have also not made any explicit assumptions about the ontological order or hierarchy of the various types of engagement which may influence their temporal relationships. For example, task analysis during the forethought phase may continue during performance, and self-reflection may overlap with performance. However, the dynamic, integrated, and contextual aspects of the model aim to highlight those interactions between facets of engagement over time. In general, research heuristics, including those regarding the aforementioned data channels and temporal granularity, are always subject to exceptions. For example, fixations are often of a social nature, and social interactions are often brief. However, in the context of measuring facets of learning engagement, it is often the case that fixations are used to examine engagement with learning content, even during collaborative tasks (Vrzakova et al., 2021). An additional limitation is that this chapter reviewed many unimodal and multimodal approaches, however we make no remarks about which of these approaches are best (due largely to conceptual and definitional differences). Moreover, although this work forwards heuristics for measure selection, we recognize that more work is needed in terms of formal review and empirical testing.

7 Concluding Thoughts

In this chapter, we introduced an expansion of the integrative model of SRL engagement (Li & Lajoie, 2021) to include emotional, behavioral, and agentic facets of engagement, based on the interrelated aspects of student engagement (Reeve, 2012). We then briefly reviewed the current conceptual, theoretical, and methodological approaches to measuring engagement, and showcased how the use of multimodal data for this work has contributed to our understanding of engagement. We extended previous literature by proposing a methodological overview to inform the research study design of future testing and validation of the IMMSE using multimodal data. Our methodological overview identifies how different modalities of measurement contribute to the measurement of engagement as it fluctuates within the different phases of SRL. We concluded our chapter with several recommendations for future research and system design.

For engagement-sensitive AIEd systems to be designed with underlying student models and adaptive scaffolding approaches, it is first imperative that the environments be able to accurately detect and infer fluctuating levels of engagement. As such, this work seeks to encourage the use of a theoretically driven model to indicate what types of data are most appropriately suited for inferences at various phases of SRL. For example, while most work has used behavioral markers as evidence of cognitive engagement (e.g., environment interactions), we show that measures of eye-tracking may be better suited for cognitive inferences (i.e., cognitive band level) while log files are the behavioral manifestations of engagement at the task level (i.e., rational band level). While both are examples of engagement, our model allows for a distinction on the type of engagement.

Measuring multidimensional facets of SRL engagement with multimodal data raise issues related to ethics, privacy, bias, transparency, and responsibility (Giannakos et al., 2022). Our model emphasizes research and training on the ethical implications of multimodal data proliferation into various facets of multimodal data including detecting, measuring, tracking, modeling, and fostering human learning with AI-based intelligent systems. As researchers we should be deeply committed to addressing ethical value conflicts that are widely known to be related to AI-based research and development including agency (consent and control), dignity (respect for persons and information systems), equity (fairness and unbiased processes), privacy (confidentiality, freedom from intrusion and interference), responsibility (of developers, users, and AI systems themselves), and trust (by users of systems and of data returned by systems). Conflicts among these values are represented through a range of practical, technical, and scientific problems including (1) who consents and does not consent to participate in research where multimodal data is critical to understand SRL engagement, (2) how much and which multimodal data is collected and from whom and where, (3) training on how to collect, analyze, interpret multimodal data, and (4) access to methods, tools, and techniques to analyze multimodal ethically and scientifically.

We argue future research testing our model fundamentally prioritizes the value of equity and fairness as a guiding principle in all our research practices, following national and international guidelines for ethical multimodal data collection, especially when considering the design of intelligent learning systems (Sharma & Giannakos, 2020). We believe that interdisciplinary researchers must be required to develop equity-focused habits of mind, which include noticing, decoding, and deconstructing machine bias and algorithmic discrimination (Cukurova et al., 2020). For example, researchers need to develop competency in strategies to mitigate AI's reification of systemic forms of social inequality (e.g., racial biases, prejudices). In addition, there are fundamental questions that may cause additional challenges that still need to be addressed by researchers. For example, what are the tradeoffs between consenting to some but not all possible multimodal data and the impacts on potential bias in data interpretation and inferences. How long should multimodal data be retained and in what forms? How is access and data sharing negotiated and coordinated between and across collaborators and academics and industry partners? How are learners made aware of what data are being collected and given options and agency (not agentic engagement but just agency as people in the world) to have voice in what is being inferred? How will explainable AI be unbiased if human researchers are using algorithms, computational models, etc., that are inherently biases because they have been developed by humans and in most cases still include the human-in-the-loop? These are some of the major challenges that multimodal data pose that will need to be addressed in order to avoid biases, prejudice, and potential abuse and misuse of multimodal data as technological advances make it easier for the ubiquitous detection, tracking, modeling of multimodal engagement data. This work serves as the base for a guide to the future direction for both researchers and instructional designers to improve the capturing and analyzing of engagement in AIEd systems using multimodal data.

Acknowledgments We wish to thank all of the current and past members of the UCF SMART Lab for their support with our ongoing research. The contributions of Roger Azevedo have been supported by several grants from the National Science Foundation (DRL#1661202, DRL#1916417, IIS#1917728, and BCS#2128684).

References

Amon, M. J., Vrzakova, H., & D'Mello, S. K. (2019). Beyond dyadic coordination: Multimodal behavioral irregularity in triads predicts facets of collaborative problem solving. *Cognitive Science, 43*(10), e12787.

Amon, M. J., Mattingly, S., Necaise, A., Mark, G., Chawla, N., & D'Mello, S. K. (2022). Flexibility versus routineness in multimodal health indicators: A sensor-based longitudinal in situ study on information workers. *ACM Transactions on Computing for Healthcare, 3*, 1. https://doi.org/10.1145/3514259

Anderson, J. R. (2002). Spanning seven orders of magnitude: A challenge for cognitive modeling. *Cognitive Science, 26*(1), 85–112.

Antonietti, A., Colombo, B., & Di Nuzzo, C. (2015). Metacognition in self-regulated multimedia learning: Integrating behavioural, psychophysiological and introspective measures. *Learning, Media and Technology, 40*(2), 187–209.

Appelhans, B. M., & Luecken, L. J. (2006). Heart rate variability as an index of regulated emotional responding. *Review of General Psychology, 10*(3), 229–240.

Appleton, J. J., Christenson, S. L., Kim, D., & Reschly, A. L. (2006). Measuring cognitive and psychological engagement: Validation of the Student Engagement Instrument. *Journal of school psychology, 44*(5), 427–445.

Appleton, J. J., Christenson, S. L., & Furlong, M. J. (2008). Student engagement with school: Critical conceptual and methodological issues of the construct. *Psychology in the Schools, 45*(5), 369–386.

Ashwin, T. S., & Guddeti, R. M. R. (2019). Unobtrusive behavioral analysis of students in classroom environments using non-verbal cues. *IEEE Access, 7*, 150693–150709.

Azevedo, R. (2015). Defining and measuring engagement and learning in science: Conceptual, theoretical, methodological, and analytical issues. *Educational Psychologist, 50*(1), 84–94.

Azevedo, R., & Gašević, D. (2019). Analyzing multimodal multichannel data about self-regulated learning with advanced learning technologies: Issues and challenges. *Computers in Human Behavior, 96*, 207–210.

Azevedo, R., Taub, M., & Mudrick, N. (2015). Think-aloud protocol analysis. In M. Spector, C. Kim, T. Johnson, W. Savenye, D. Ifenthaler, & G. Del Rio (Eds.), *The SAGE encyclopedia of educational technology* (pp. 763–766). SAGE.

Azevedo, R., Taub, M., & Mudrick, N. V. (2017). Understanding and reasoning about real-time cognitive, affective, and metacognitive processes to foster self-regulation with advanced learning technologies. In D. H. Schunk & J. A. Greene (Eds.), *Handbook of self-regulation of learning and performance* (pp. 254–270). Routledge.

Azevedo, R., Mudrick, N. V., Taub, M., & Bradbury, A. E. (2019). Self-regulation in computer-assisted learning systems. In J. Dunlosky & K. A. Rawson (Eds.), *The Cambridge handbook of cognition and education* (pp. 587–618). Cambridge University Press. https://doi.org/10.1017/9781108235631.024

Azevedo, R., Bouchet, F., Duffy, M., Harley, J., Taub, M., Trevors, G., et al. (2022). Lessons learned and future directions of MetaTutor: Leveraging multichannel data to scaffold self-regulated learning with an intelligent tutoring system. *Frontiers in Psychology, 13*.

Azevedo, R., & Wiedbusch, M. (2023). Theories of metacognition and pedagogy applied to AIED systems. In Handbook of Artificial Intelligence in Education (pp. 45–67). Edward Elgar Publishing.

Baker, R. S., Corbett, A. T., Roll, I., Koedinger, K. R., Aleven, V., Cocea, M., et al. (2013). Modeling and studying gaming the system with educational data mining. In R. Azevedo & V. Aleven (Eds.), *International handbook of metacognition and learning technologies* (pp. 97–115). Springer.

Baker, R. S., Ocumpaugh, J. L., & Andres, J. M. A. L. (in press). BROMP quantitative field observations: A review. In R. Feldman (Ed.), *Learning science: Theory, research, and practice*. McGraw-Hill.

Bernacki, M. L., Byrnes, J. P., & Cromley, J. G. (2012). The effects of achievement goals and self-regulated learning behaviors on reading comprehension in technology-enhanced learning environments. *Contemporary Educational Psychology, 37*(2), 148–161.

Bidwell, J., & Fuchs, H. (2011). Classroom analytics: Measuring student engagement with automated gaze tracking. *Behavior Research Methods, 49*(113).

Bixler, R., & D'Mello, S. (2016). Automatic gaze-based user-independent detection of mind wandering during computerized reading. *User Modeling and User-Adapted Interaction, 26*(1), 33–68.

Boekaerts, M. (2016). Engagement as an inherent aspect of the learning process. *Learning and Instruction, 43*, 76–83.

Boucheix, J. M., Lowe, R. K., Putri, D. K., & Groff, J. (2013). Cueing animations: Dynamic signaling aids information extraction and comprehension. *Learning and Instruction, 25*, 71–84.

Carter, B. T., & Luke, S. G. (2020). Best practices in eye tracking research. *International Journal of Psychophysiology, 155*, 49–62.

Chapman, C. M., Deane, K. L., Harré, N., Courtney, M. G., & Moore, J. (2017). Engagement and mentor support as drivers of social development in the project K youth development program. *Journal of Youth and Adolescence, 46*(3), 644–655.

Cheon, S. H., Reeve, J., & Ntoumanis, N. (2018). A needs-supportive intervention to help PE teachers enhance students' prosocial behavior and diminish antisocial behavior. *Psychology of Sport and Exercise, 35*, 74–88.

Connell, J. P., & Wellborn, J. G. (1991). Competence, autonomy, and relatedness: A motivational analysis of self-system processes. In M. R. Gunnar & L. A. Sroufe (Eds.), *Self processes and development* (pp. 43–77). Lawrence Erlbaum Associates, Inc.

Corno, L., & Mandinach, E. B. (1983). The role of cognitive engagement in classroom learning and motivation. *Educational Psychologist, 18*(2), 88–108.

Craig, S. D., D'Mello, S., Witherspoon, A., & Graesser, A. (2008). Emote aloud during learning with AutoTutor: Applying the facial action coding system to cognitive–affective states during learning. *Cognition and Emotion, 22*(5), 777–788.

Cukurova, M., Giannakos, M., & Martinez-Maldonado, R. (2020). The promise and challenges of multimodal learning analytics. *British Journal of Educational Technology, 51*(5), 1441–1449. https://doi.org/10.1111/bjet.13015

D'Mello, S. K., & Mills, C. S. (2021). Mind wandering during reading: An interdisciplinary and integrative review of psychological, computing, and intervention research and theory. *Language and Linguistics Compass, 15*(4), e12412.

D'Mello, S. K., Craig, S. D., Sullins, J., & Graesser, A. C. (2006). Predicting affective states expressed through an emote-aloud procedure from AutoTutor's mixed-initiative dialogue. *International Journal of Artificial Intelligence in Education, 16*(1), 3–28.

D'Mello, S., Olney, A., Williams, C., & Hays, P. (2012). Gaze tutor: A gaze-reactive intelligent tutoring system. *International Journal of Human-Computer Studies, 70*(5), 377–398.

D'Mello, S. K., Dieterle, E., & Duckworth, A. (2017). Advanced, Analytic, Automated (AAA) measurement of engagement during learning. *Educational Psychologist, 52*(2), 104–123.

Darvishi, A., Khosravi, H., Sadiq, S., & Weber, B. (2021). Neurophysiological measurements in higher education: A systematic literature review. *International Journal of Artificial Intelligence in Education*, 1–41.

Dent, A. L., & Koenka, A. C. (2016). The relation between self-regulated learning and academic achievement across childhood and adolescence: A meta-analysis. *Educational Psychology Review, 28*, 425–474.

Dever, D. A., Azevedo, R., Cloude, E. B., & Wiedbusch, M. (2020). The impact of autonomy and types of informational text presentations in game-based environments on learning: Converging multi-channel processes data and learning outcomes. *International Journal of Artificial Intelligence in Education, 30*(4), 581–615.

Dewan, M. A. A., Murshed, M., & Lin, F. (2019). Engagement detection in online learning: A review. *Smart Learning. Environments., 6*, 1. https://doi.org/10.1186/s40561-018-0080-z

Di Lascio, E., Gashi, S., & Santini, S. (2018). Unobtrusive assessment of students' emotional engagement during lectures using electrodermal activity sensors. *Proceedings of the ACM on Interactive, Mobile, Wearable and Ubiquitous Technologies, 2*(3), 1–21.

Dubovi, I. (2022). Cognitive and emotional engagement while learning with VR: The perspective of multimodal methodology. *Computers & Education, 183*, 104495.

Duchowski, A. (2007). Eye Tracking Techniques. In: Eye Tracking Methodology. Springer, London. https://doi.org/10.1007/978-1-84628-609-4_5

Duffy, M. C., & Azevedo, R. (2015). Motivation matters: Interactions between achievement goals and agent scaffolding for self-regulated learning within an intelligent tutoring system. *Computers in Human Behavior, 52*, 338–348.

Ekman, P., & Friesen, W. V. (1978). Facial action coding system. *Environmental Psychology & Nonverbal Behavior.*

Ericsson, K. A., & Simon, H. A. (1984). *Protocol analysis: Verbal reports as data*. The MIT Press.

Fredricks, J. A., Blumenfeld, P. C., & Paris, A. H. (2004). School engagement: Potential of the concept, state of the evidence. *Review of educational research, 74*(1), 59–109.

Fredricks, J. A., & McColskey, W. (2012). The measurement of student engagement: A comparative analysis of various methods and student self-report instruments. In S. L. Christenson, A. L. Reschly, & C. Wylie (Eds.), *Handbook of research on student engagement* (pp. 763–782). Springer.

Fredricks, J. A., Filsecker, M., & Lawson, M. A. (2016). Student engagement, context, and adjustment: Addressing definitional, measurement, and methodological issues. *Learning and Instruction, 43*, 1–4.

Fredricks, J., Hofkens, T., & Wang, M. (2019a). Addressing the challenge of measuring student engagement. In K. Renninger & S. Hidi (Eds.), *The Cambridge handbook of motivation and learning* (pp. 689–712). Cambridge University Press. https://doi.org/10.1017/9781316823279.029

Fredricks, J. A., Reschly, A. L., & Christenson, S. L. (2019b). Interventions for student engagement: Overview and state of the field. In J. A. Fredricks, A. L. Reschly, & S. Christenson (Eds.), *Handbook of student engagement interventions* (pp. 1–11). Academic Press. https://doi.org/10.1016/C2016-0-04519-9

Freeman, F. G., Mikulka, P. J., Prinzel, L. J., & Scerbo, M. W. (1999). Evaluation of an adaptive automation system using three EEG indices with a visual tracking task. *Biological Psychology, 50*(1), 61–76.

Gao, N., Shao, W., Rahaman, M. S., & Salim, F. D. (2020). n-gage: Predicting in-class emotional, behavioural and cognitive engagement in the wild. *Proceedings of the ACM on Interactive, Mobile, Wearable and Ubiquitous Technologies, 4*(3), 1–26.

Gevins, A., & Smith, M. E. (2008). Electroencephalography (EEG) in neuroergonomics. In R. Parasuraman & M. Rizzo (Eds.), *Neuroergonomics: The brain at work* (pp. 15–31). Oxford University Press.

Giannakos, M., Spikol, D., Di Mitri, D., Sharma, K., Ochoa, X., & Hammad, R. (Eds.). (2022). *The multimodal learning analytics handbook*. Springer.

Gobert, J. D., Baker, R. S., & Wixon, M. B. (2015). Operationalizing and detecting disengagement within online science microworlds. *Educational Psychologist, 50*(1), 43–57.

Grafsgaard, J., Wiggins, J. B., Boyer, K. E., Wiebe, E. N., & Lester, J. (2013). Automatically recognizing facial expression: Predicting engagement and frustration. In *Proceedings of the international conference on Educational data mining*.

Greene, J. A., Oswald, C. A., & Pomerantz, J. (2015). Predictors of retention and achievement in a massive open online course. *American Educational Research Journal, 52*(5), 925–955.

Greene, B. A. (2015). Measuring cognitive engagement with self-report scales: Reflections from over 20 years of research. *Educational Psychologist, 50*(1), 14–30.

Greene, J. A., & Azevedo, R. (2010). The measurement of learners' self-regulated cognitive and metacognitive processes while using computer-based learning environments. *Educational Psychologist, 45*(4), 203–209.

Haerens, L., Aelterman, N., Vansteenkiste, M., Soenens, B., & Van Petegem, S. (2015). Do perceived autonomy-supportive and controlling teaching relate to physical education students' motivational experiences through unique pathways? Distinguishing between the bright and dark side of motivation. *Psychology of Sport and Exercise, 16*, 26–36.

Harley, J. M., Pekrun, R., Taxer, J. L., & Gross, J. J. (2019). Emotion regulation in achievement situations: An integrated model. *Educational Psychologist, 54*(2), 106–126. https://doi.org/10.1080/00461520.2019.1587297

Hassib, M., Khamis, M., Friedl, S., Schneegass, S., & Alt, F. (2017a). Brainatwork: Logging cognitive engagement and tasks in the workplace using electroencephalography. In *Proceedings of the 16th international conference on mobile and ubiquitous multimedia* (pp. 305–310).

Hassib, M., Schneegass, S., Eiglsperger, P., Henze, N., Schmidt, A., & Alt, F. (2017b). EngageMeter: A system for implicit audience engagement sensing using electroencephalography. In *Proceedings of the 2017 Chi conference on human factors in computing systems* (pp. 5114–5119).

Haubert, A., Walsh, M., Boyd, R., Morris, M., Wiedbusch, M., Krusmark, M., & Gunzelmann, G. (2018). Relationship of event-related potentials to the vigilance decrement. *Frontiers in Psychology, 9*, 237.

Helme, S., & Clarke, D. (2001). Identifying cognitive engagement in the mathematics classroom. *Mathematics Education Research Journal, 13*(2), 133–153.

Henrie, C. R., Halverson, L. R., & Graham, C. R. (2015). Measuring student engagement in technology-mediated learning: A review. *Computers & Education, 90*, 36–53.

Huang, J., Yu, C., Wang, Y., Zhao, Y., Liu, S., Mo, C., … & Shi, Y. (2014, April). FOCUS: enhancing children's engagement in reading by using contextual BCI training sessions. In *Proceedings of the SIGCHI conference on human factors in computing systems* (pp. 1905–1908).

Ireland, M. E., & Henderson, M. D. (2014). Language style matching, engagement, and impasse in negotiations. *Negotiation and Conflict Management Research, 7*(1), 1–16.

Jang, H., Kim, E. J., & Reeve, J. (2016). Why students become more engaged or more disengaged during the semester: A self-determination theory dual-process model. *Learning and Instruction, 43*, 27–38.

Järvelä, S., & Bannert, M. (2021). Temporal and adaptive processes of regulated learning – What can multimodal data tell? *Learning and Instruction, 72*, 101268.

Järvelä, S., Veermans, M., & Leinonen, P. (2008). Investigating student engagement in computer-supported inquiry: A process-oriented analysis. *Social Psychology of Education, 11*(3), 299–322.

Kew, S. N., & Tasir, Z. (2021). Analyzing students' cognitive engagement in e-learning discussion forums through content analysis. *Knowledge Management & E-Learning: An International Journal, 13*(1), 39–57.

Killingsworth, M. A., & Gilbert, D. T. (2010). A wandering mind is an unhappy mind. *Science, 330*(6006), 932–932.

Klimesch, W. (2012). Alpha-band oscillations, attention, and controlled access to stored information. *Trends in Cognitive Sciences, 16*(12), 606–617.

Kruger, J. L., Hefer, E., & Matthew, G. (2014). Attention distribution and cognitive load in a subtitled academic lecture: L1 vs. L2. *Journal of Eye Movement Research, 7*(5).

Lee, J., Song, H. D., & Hong, A. J. (2019). Exploring factors, and indicators for measuring students' sustainable engagement in e-learning. *Sustainability, 11*(4), 985.

Li, S., & Lajoie, S. P. (2021). Cognitive engagement in self-regulated learning: An integrative model. *European Journal of Psychology of Education*, 1–20.

Li, S., Zheng, J., & Lajoie, S. P. (2020). The relationship between cognitive engagement and students' performance in a simulation-based training environment: An information-processing perspective. *Interactive Learning Environments*, 1–14.

Li, S., Lajoie, S. P., Zheng, J., Wu, H., & Cheng, H. (2021). Automated detection of cognitive engagement to inform the art of staying engaged in problem-solving. *Computers & Education, 163*, 104114.

Lin, F. R., & Kao, C. M. (2018). Mental effort detection using EEG data in E-learning contexts. *Computers & Education, 122*, 63–79.

Mandernach, B. J. (2015). Assessment of student engagement in higher education: A synthesis of literature and assessment tools. *International Journal of Learning, Teaching and Educational Research, 12*(2).

Molenaar, I., de Mooij, S., Azevedo, R., Bannertd, M., Järveläe, S., & Gaševićf, D. (2022). Measuring self-regulated learning and the role of AI: Five years of research using multimodal multichannel data. *Computers in Human Behavior*, 107540.

Miller, B. W. (2015). Using reading times and eye-movements to measure cognitive engagement. *Educational Psychologist, 50*(1), 31–42.

Miller, R. B., Greene, B. A., Montalvo, G. P., Ravindran, B., & Nichols, J. D. (1996). Engagement in academic work: The role of learning goals, future consequences, pleasing others, and perceived ability. *Contemporary Educational Psychology, 21*(4), 388–422.

Newell, A. (1994). *Unified theories of cognition*. Harvard University Press.

Nuamah, J. K., & Seong, Y. (2018). Support vector machine (SVM) classification of cognitive tasks based on electroencephalography (EEG) engagement index. *Brain-Computer Interfaces, 5*(1), 1–12.

Oshima, J., & Hoppe, H. U. (2021). Finding meaning in log-file data. In U. Cress, C. Rosé, A. F. Wise, & J. Oshima (Eds.), *International handbook of computer-supported collaborative learning* (pp. 569–584). Springer.

Panadero, E. (2017). A review of self-regulated learning: Six models and four directions for research. *Frontiers in Psychology, 8*, 422.

Papamitsiou, Z., Pappas, I. O., Sharma, K., & Giannakos, M. N. (2020). Utilizing multimodal data through fsQCA to explain engagement in adaptive learning. *IEEE Transactions on Learning Technologies, 13*(4), 689–703.

Penttinen, M., Anto, E., & Mikkilä-Erdmann, M. (2013). Conceptual change, text comprehension and eye movements during reading. *Research in Science Education, 43*(4), 1407–1434.

Perry, N. E. (2002). Introduction: Using qualitative methods to enrich understandings of self-regulated learning. *Educational Psychologist, 37*(1), 1–3.

Perry, N. E., & Winne, P. H. (2006). Learning from learning kits: gStudy traces of students' self-regulated engagements with computerized content. *Educational Psychology Review, 18*(3), 211–228.

Pope, A. T., Bogart, E. H., & Bartolome, D. S. (1995). Biocybernetic system evaluates indices of operator engagement in automated task. *Biological Psychology, 40*(1–2), 187–195.

Posada-Quintero, H. F., & Chon, K. H. (2019). Innovations in electrodermal activity data collection and signal processing: A systematic review. *Sensors, 20*, 1–18.

Ramachandran, A., Huang, C. M., Gartland, E., & Scassellati, B. (2018, February). Thinking aloud with a tutoring robot to enhance learning. In *Proceedings of the 2018 ACM/IEEE international conference on human-robot interaction* (pp. 59–68).

Reeve, J. (2012). A self-determination theory perspective on student engagement. In S. Christenson, A. L. Reschly, & C. Wylie (Eds.), *Handbook of research on student engagement* (pp. 149–172). Springer.

Reeve, J. (2013). How students create motivationally supportive learning environments for themselves: The concept of agentic engagement. *Journal of Educational Psychology, 105*(3), 579.

Reeve, J., Cheon, S. H., & Jang, H. R. (2019). A teacher-focused intervention to enhance students' classroom engagement. In J. Fredricks, A. L. Reschly, & S. Christenson (Eds.), *Handbook of student engagement interventions* (pp. 87–102). Academic Press.

Renninger, K. A., & Bachrach, J. E. (2015). Studying triggers for interest and engagement using observational methods. *Educational Psychologist, 50*(1), 58–69.

Reschly, A. L., & Christenson, S. L. (2012). Jingle, jangle, and conceptual haziness: Evolution and future directions of the engagement construct. In A. L. Christenson, A. L. Reschly, & C. Wylie (Eds.), *Handbook of research on student engagement* (pp. 3–19). Springer.

Salmela-Aro, K., Moeller, J., Schneider, B., Spicer, J., & Lavonen, J. (2016). Integrating the light and dark sides of student engagement using person-oriented and situation-specific approaches. *Learning and Instruction, 43*, 61–70.

Schunk, D. H., & Greene, J. A. (2017). Historical, contemporary, and future perspectives on self-regulated learning and performance. In D. H. Schunk & J. A. Greene (Eds.), *Handbook of self-regulation of learning and performance* (pp. 1–15). Routledge.

Sharma, K., & Giannakos, M. (2020). Multimodal data capabilities for learning: What can multimodal data tell us about learning? *British Journal of Educational Technology, 51*(5), 1450–1484. https://doi.org/10.1111/bjet.12993

Sharma, P., Joshi, S., Gautam, S., Maharjan, S., Filipe, V., & Reis, M. J. (2019). Student engagement detection using emotion analysis, eye tracking and head movement with machine learning. *arXiv:1909.12913*. https://doi.org/10.48550/arXiv.1909.12913

Shukor, N. A., Tasir, Z., Van der Meijden, H., & Harun, J. (2014). A predictive model to evaluate students' cognitive engagement in online learning. *Procedia-Social and Behavioral Sciences, 116*, 4844–4853.

Sinatra, G. M., Heddy, B. C., & Lombardi, D. (2015). The challenges of defining and measuring student engagement in science. *Educational Psychologist, 50*(1), 1–13.

Skinner, E. A., Kindermann, T. A., & Furrer, C. J. (2009). A motivational perspective on engagement and disaffection: Conceptualization and assessment of children's behavioral and emotional participation in academic activities in the classroom. *Educational and Psychological Measurement, 69*(3), 493–525.

Taub, M., & Azevedo, R. (2019). How does prior knowledge influence eye fixations and sequences of cognitive and metacognitive SRL processes during learning with an intelligent tutoring system? *International Journal of Artificial Intelligence in Education, 29*(1), 1–28.

Taub, M., Sawyer, R., Smith, A., Rowe, J., Azevedo, R., & Lester, J. (2020). The agency effect: The impact of student agency on learning, emotions, and problem-solving behaviors in a game-based learning environment. *Computers & Education, 147*, 103781.

Tausczik, Y. R., & Pennebaker, J. W. (2010). The psychological meaning of words: LIWC and computerized text analysis methods. *Journal of Language and Social Psychology, 29*(1), 24–54.

Terriault, P., Kozanitis, A., & Farand, P. (2021). Use of electrodermal wristbands to measure students' cognitive engagement in the classroom. In *Proceedings of the Canadian Engineering Education Association (CEEA)*.

van der Graaf, J., Lim, L., Fan, Y., Kilgour, J., Moore, J., Gašević, D., et al. (2022). The dynamics between self-regulated learning and learning outcomes: An exploratory approach and implications. *Metacognition and Learning*, 1–27.

Van Gog, T., & Jarodzka, H. (2013). Eye tracking as a tool to study and enhance cognitive and metacognitive processes in computer-based learning environments. *International handbook of metacognition and learning technologies*, 143–156.

Veenman, M. V. J., & van Cleef, D. (2019). Measuring metacognitive skills for mathematics: Students' self-reports versus on-line assessment methods. *ZDM Mathematics Education, 51*, 691–701. https://doi.org/10.1007/s11858-018-1006-5

Verdière, K. J., Roy, R. N., & Dehais, F. (2018). Detecting pilot's engagement using fNIRS connectivity features in an automated vs. manual landing scenario. *Frontiers in human neuroscience, 12*, 6.

Vongkulluksn, V. W., Lu, L., Nelson, M. J., & Xie, K. (2022). Cognitive engagement with technology scale: A validation study. *Educational Rechnology Research and Development, 70*, 1–27.

Vrzakova, H., Amon, M. J., & D'Mello, S. K. (2021). Looking for a deal! Visual social attention during negotiations via mixed media videoconferencing. *Proceedings of the Association for Computing Machinery: Computer Supported Cooperative Work (CSCW), 4*, 1–35. https://doi.org/10.1145/3434169

Wang, M. T., & Fredricks, J. A. (2014). The reciprocal links between school engagement, youth problem behaviors, and school dropout during adolescence. *Child Development, 85*(2), 722–737.

Wang, Y., Kotha, A., Hong, P. H., & Qiu, M. (2020, August). Automated student engagement monitoring and evaluation during learning in the wild. In *2020 7th IEEE international conference on cyber security and cloud computing (CSCloud)/2020 6th IEEE international conference on edge computing and scalable cloud (EdgeCom)* (pp. 270–275). IEEE.

West, R. L., & MacDougall, K. (2014). The macro-architecture hypothesis: Modifying Newell's system levels to include macro-cognition. *Biologically Inspired Cognitive Architectures, 8*, 140–149.

Whitehill, J., Serpell, Z., Lin, Y. C., Foster, A., & Movellan, J. R. (2014). The faces of engagement: Automatic recognition of student engagement from facial expressions. *IEEE Transactions on Affective Computing, 5*(1), 86–98.

Wiedbusch, M. D., & Azevedo, R. (2020). Modeling metacomprehension monitoring accuracy with eye gaze on informational content in a multimedia learning environment. In *ACM symposium on eye tracking research and applications* (pp. 1–9).

Winne, P. H., Jamieson-Noel, D., & Muis, K. (2002). Methodological issues and advances in researching tactics, strategies, and self-regulated learning. Advances in motivation and achievement. *New Directions in Measures and Methods, 12*, 121–155.

Wolters, C. A. (2004). Advancing achievement goal theory: Using goal structures and goal orientations to predict students' motivation, cognition, and achievement. *Journal of educational psychology, 96*(2), 236.

Wolters, C. A., & Taylor, D. J. (2012). A self-regulated learning perspective on student engagement. In A. L. Christenson, A. L. Reschly, & C. Wylie (Eds.), *Handbook of research on student engagement* (pp. 635–651). Springer.

Xie, K., Heddy, B. C., & Vongkulluksn, V. W. (2019). Examining engagement in context using experience-sampling method with mobile technology. *Contemporary Educational Psychology, 59*, 101788.

Zimmerman, B. J. (2000). Self-efficacy: An essential motive to learn. *Contemporary Educational Psychology, 25*(1), 82–91.

Chapter 11
Roles for Information in Trace Data Used to Model Self-Regulated Learning

Philip H. Winne

Abstract When researchers use software and other technologies to gather data about learning, an operational definition details what to record about timestamped learning events as a learner engages with information, e.g., selecting text in a webpage or tagging selections to index them. Theory assigns meaning to such operational definitions: (a) selecting text signals metacognitive monitoring; (b) tagging reveals properties the learner monitors as descriptive of selections, e.g., interesting, to investigate; (c) the learner ascribes utility to effort spent to select and tag. Prevailing approaches to analyzing trace data examine events in terms of presence/absence, frequency, contingency, and pattern. For example, does the learner metacognitively monitor? How many times? If the learner tags information "interesting," does the learner contingently search for supplementary information? Properties of the information on which learners operate are underappreciated in analyses of trace data. What features of information lead a learner to: rehearse it vs. not; … tag it important vs. interesting vs. to investigate? … annotate it vs. search for supplemental material? … bin it, e.g., very difficult or not worth effort to learn? This chapter explores roles for information as information that can enrich trace data describing learning events. For example, can information a learner tags imply prior knowledge? Do tags signal mastery vs. performance goal orientation? Attending to information as information expands views about trace data and their uses in learning analytics and researching self-regulated learning.

Keywords Trace data · Learning events · Self-regulated learning · Learning analytics

---

P. H. Winne (✉)
Simon Fraser University, Burnaby, BC, Canada
e-mail: winne@sfu.ca

V. Kovanovic et al. (eds.), *Unobtrusive Observations of Learning in Digital Environments*, Advances in Analytics for Learning and Teaching,
https://doi.org/10.1007/978-3-031-30992-2_11

1 Introduction

Research literatures about online learning, self-regulated learning (SRL), learning science, and learning analytics often refer to and analyze processes involving cognition, metacognition, and motivation. Processes label operations learners are theorized to apply to information (Winne, 2018, in press). For example, rehearsing is one cognitive operation. It reproduces specific information in working memory, theoretically with near perfect accuracy. Monitoring is another cognitive operation. It produces a list recording matches or a profile comparing properties of a "target" chunk of information, an object, to properties of a "standard" chunk of information. Monitoring can be a cognitive or a metacognitive operation depending on whether information monitored *is* the topic of a task – What are steps in graphing a linear equation? – or information *about* the topic of a task – Do I feel more confident graphing a linear equation using method A or method B? A motivational operation is choosing among options. For example, studying art history to develop knowledge for its own sake is a choice among reasons for studying. This choice represents mastery goal orientation. Or, the choice of reasons for studying art history may be to prepare to demonstrate expertise to others. Choosing this reason to justify behavior represents performance goal orientation.

Operationally defining operations in learning is challenging. For example, a learner surveying a webpage to identify source material to use in a term paper may judge (monitor) the content is uninteresting. Or, the learner may judge this source is helpful because text descriptions of complex systems or principles are translated as diagrams. How can judgments like these be observed? A learning scientist may ask the learner to talk aloud while working, hoping the learner reports each learning event precisely, fully, and reliably. Some researchers have used facial recognition technologies coupled with systems tracking eye gaze to assemble a signal they interpret as the learner reaching a judgment like this.

A third approach is to operationally define trace data. Trace data are typically recorded in software logs when learners use software features on-the-fly. Instances or patterns of trace data are theorized to correspond to fundamental operations and patterns of operations that manipulate information (Winne, 2020a). For example, learners may select (monitor) and tag (assemble) text *interesting*. Text not tagged is inferred to have been monitored as uninteresting per se or not sufficiently interesting or otherwise of value to be selected and tagged. Or learners may annotate a diagram using a schema operationalized as a structured note form in which distinct labels for each of several fields prompt the learner to describe key features of a system, their functions, contingencies, and other properties recognized in the diagram.

Researchers are actively exploring how to operationally define operations learners engage during work on assigned and self-chosen tasks. The vast majority of this work addresses a basic question: What did the learner do? Answers often take form as an account of singular events or patterns relating learning events. Learning events can be ordered across a timeline of their occurrences. Both relatively simple and rather sophisticated methodologies – graph theory (Winne et al., 1994) and process mining (Saint et al., 2021), respectively – are available to characterize contingencies and patterns of learning events. In this chapter, I develop a case that approaches like these give too little attention to the information operated on in a learning event. In effect, those methods describe "empty" learning events. Theorizing how learning events relate to knowledge a learner develops (or doesn't), motivation guiding a learner's choices, and affect a learner experiences requires incorporating information in accounts of learning events because self-regulating learners select operations they apply according to content and properties of information. To approach clarity needed to observe and measure that information, a first step is describing how a learning event can be modeled.

2 Learning Events

The literatures mentioned earlier describe learning events as operations or processes. Operations manipulate information. I posit a set of basic operations referenced by the first-letter mnemonic SMART: searching, monitoring, assembling, rehearsing, and translating (e.g., Winne, 2018, in press). Table 11.1 provides definitions and examples. As entries in Table 11.1 describe, operations like the SMART set inherently require inputs and generate products. Inputs may be elemental propositions describing any topic, including feelings and reasons for engaging in behavior, i.e., motivations. (See Renninger & Hidi, 2019 for a compendium of motivation theories positing reasons for behavior.) Inputs also can be complex structures of information, such as a graph contrasting changes in energy levels across the lifespan of a catalytically assisted chemical reaction as contrasted to that reaction without the catalyst. Without information inputs, there would be no "content" on which to operate.

Notably, operations always are carried out in the context of surrounding conditions which may bear on how a learner regulates operations. Conditions can be external to the learner, such as whether peers are nearby to observe, or that time allowed for executing a task is nearly expired. Conditions also can be internal to the learner, such as enduring motivations, prior knowledge encoded in long-term memory and expectations the learner forecasts about standards by which a product will be evaluated. An important class of internal conditions not addressed further in this chapter but not to be forgotten are individual differences such as working memory span.

When operations are executed on information and a product is generated, the learner's state is updated. The updating of states marks a learning event. Having generated a product, the learner is now in position to monitor its properties in relation to standards for work that generated that product(s), to assemble an attribution describing that result and assemble a feeling with that information complex. Monitoring those inputs and assembling those accounts defines another learning event. For example, did work to translate the symbolic expression $y = 2x + 3$ into graphic form proceed straightforwardly, step-by-step, or were retreats necessary to

Table 11.1 The SMART operations

Operation	Input	Product	Example
Searching	Information active in working memory. This includes perceptions about features in the external environment and neighborhoods in the network of long-term memory	Information elsewhere in long-term memory becomes activated in working memory because network paths connect that information to inputs	Sodium has the chemical symbol …?
Monitoring	A list or configuration (e.g., schema, step-by-step procedure) of standards for judging an information input or a product of thought or behavior	Classification (yes/no; multicategory) or rating of a target according to whether or how well its properties correspond to standards	A zebra displays each defining characteristic of a mammal
Assembling	Two or more units of information (e.g., propositions, chunks, instantiated schemas) active in working memory	A relational property describing the union or intersection of the units of information	If the temperature of water at standard pressure exceeds 100 °C, then its state changes from liquid to gas
Rehearsing	One unit of information active in working memory	A (near perfect) reproduction of active information in working memory	Mentally repeating an assembly relating the term *deciduous* to its definition
Translating	A unit of information active in working memory	A re-presentation of the input in a changed form that preserves core meaning, and possibly introduces new information	A paraphrase A graph of $y = 2x + 3$

correct errors? Is a correct graph attributed to dedicated effort or “dumb luck”? Does that attribution engender a feeling of efficacy or anxiety about similar future tasks? Each of these products arises in a context of previously updated information – internal and external conditions – as the task unfolds. Those conditions are examined by the self-regulating learner to make next choices about possible operations, operations actually applied, products each set of operations can generate, and evaluations of those products in reference to standards. A first letter mnemonic COPES – conditions, operations, products, evaluations, and standards – assembles these information topics as a unit.

Elsewhere, I model a bundle of internal conditions contributing to a learner’s decision-making policy about whether and how to engage tasks. Making choices about tasks enacts motivation. One facet of motivation is attributions, reasons a learner constructs to explain evaluations of products (Weiner, 2010). Efficacy expectations are the learner’s predictions about the degree to which current knowledge and skills are available to succeed at a task. Efficacy expectations are informed by standards that characterize a high-quality product. Outcome expectations are the learner's perceptions about what product will result if particular operations are executed, and what are the properties of that product. Efficacy and outcome expectations are pillars in Bandura's model of social learning (Bandura, 1997). Incentives are values the learner associates with COPES aspects of tasks as well as emotions arising from attributions (Weiner, 2010). Based on these perceptions about conditions, the learner constructs a utility judgment for each task: What is the balance of costs relative to benefits if a task is engaged by applying particular operations under present conditions when particular standards apply? AEIOU – attribution, efficacy expectation, incentive, outcome expectation, and utility – is a convenient first-letter mnemonic assembling these internal products of cognition into one unit (Winne, 2022).

These three schemas jointly characterize features of learning events. The set of SMART operations distinguishes operations for processing and creating information by inputs and products. COPES identifies facets of information describing a task in which operations, SMARTs, are executed. Information the learner produces in the form of AEIOU assembles motivation and affect with COPES.

States are point-in-time snapshots. A state is stable for a brief instant when it materializes, then it is replaced by the next state as subsequent operations generate new products. That transition marks a learning event. Learning events arising across the timeline of a task represent learning as a dynamically connected series of autoregressive states.

2.1 Modeling One Learning Event: If-Then-Else

I borrowed from other disciplines, especially computer science, to model learning as a sequence of If-Then-Else productions (e.g., Winne, 2018, in press). If collects conditions, the amalgam of external factors under which a learner may engage a task plus internal conditions integrated by the AEOIU model. Depending on the profile or constitution of Ifs, the learner Then executes one operation or a strategic pattern of operations. Should conditions be configured otherwise, then Else some other operation(s) are selected. For example, a learner encountering a technical term formatted in italics (If) regularly selects and tags it for review (Then) excepting (Else) terms which the learner already knows well.

The If-Then-Else model spans time by bridging the transition from a preceding state, If generated by monitoring information, to a subsequent state, an information product generated by Then or Else. How a learner chooses to learn – to self-regulate learning – is conditional on Ifs. Modeling learning events requires examining sequences of If-Then-Else events that modulate in response to varying Ifs. Modeling and analyzing SRL event data is dynamic because each event updates conditions characterizing the next moment in time.

It merits pointing out this model emphasizes the learner is in full control. The learner perceives states and chooses how to behave. This includes how to think, which operations are applied to what information. While observers and even learners may interpret choice is removed when learning is habitual (automatically engaged with apparently no deliberation or apparent draw on resources of working memory), that is a false proposition. SRL is ubiquitous but its forms vary based on information the learner processes, potentially moderated by external conditions (Winne, 1995). Automated routines encapsulate SRL in ways that bury inside automated productions a learner's choices about learning. Observers and even learners can be unaware of complex cognition (Vatansever et al., 2017). Choice was front-and-center, however, when such routines were first created and along the way leading to automated status.

3 Information Is the Subject of Operations

Every facet in each of the COPES, SMART, and AEIOU models is centered on information a learner attends to and uses in the course of SRL. In the case of SMART operations and strategic assemblies of them, more commonly called learning strategies, the information referred to is steps in a procedure, a script. What is the role of information in SRL, specifically, in motivation, cognition, and metacognition? The next three sections illustrate answers to this question, laying groundwork for this proposition: When accounts of SRL are limited to occurrences, frequencies, or patterns of operations (processes), those accounts cannot represent enough of the story of SRL.

3.1 Motivation

A learner's motivation has an explicit topic. Learners are curious about certain subjects, appreciate feedback with particular properties, or are anxious about a specific social event. Motivation is also situationally anchored. For example, a learner participating in a think-aloud protocol might remark, "I think I can solve *this* problem but I need to be careful" (emphasis added). This utterance is referenced to specific external conditions the learner perceives in this moment. This information lies alongside memories the learner samples from their experiential history. Sampling is influenced by the learner's perceptions about current external conditions, such as whether an answer key is available which would afford the option to select a strategy of working backward. Jointly, these conditions figure into the learner's choice about how to proceed. Every self-report questionnaire I have examined reflects the situationality of motivation. Instructions to respondents set boundaries on the situation they are asked to keep in mind as they respond to questionnaire items. For example, a questionnaire's instructions may advise the learner to consider "this course" or the discipline of "science" when rating motivation about the incentive to score higher on achievement measures than classmates (performance goal orientation) or as a measure of subject matter mastery (mastery goal orientation).

Self-report data are problematic (Winne, 2020b), in part because humans have fallible and biased memories of past experience, and because they may unintentionally bias perceptions about current states and events. Modern technologies such as software logging, and facial recognition and eye tracking systems may improve data about motivation. For example, clickstream logging can identify whether a learner visited an assigned webpage, and eye tracking data can confirm whether a learner's gaze oscillated several times between text describing a complex relation, such as activation energy in a catalytic reaction, and a figure translating textual information about that relation (e.g., see Fig. 12.19 at https://openstax.org/books/chemistry-2e/pages/12-7-catalysis; Flowers et al., 2019). These online data can lend support to inferences about a learner's rating of motivation described by a questionnaire item about utility of a learning tactic: "Do you analyze diagrams and graphs to build understanding when you study?" But validity is still in some jeopardy. Data gathered online then coupled with the self-report datum do not reveal whether the learner analyzed information. Motivation is present, but motivation about what topic and motivation to engage in what particular cognition? To confirm the learner analyzed information, data about information input to and produced by analytic thinking is needed.

I offer this axiom regarding motivation: Behavior is motivated. Put another way, excepting for autonomic and automated responses to information states – e.g., reducing blinking rate under cognitive load (e.g., Dubovi, 2022), modulating reading pace according to punctuation (Chung & Bidelman, 2022) – learners (and people, in general) behave as they do because they deliberately reason to reach judgments about which behavior is preferred. People are rational but their rationality is rooted in *idiosyncratic* reasons and *personal* logic. Consequently, a learner's reasons and logic for motivated behavior may not correspond to norms or an instructor's goals. Learners may appear irrational from others' points of view.

The axiom that behavior is motivated stimulates extending the analysis of thinking as a behavior. The network of information that is long-term memory propagates activation across nodes of information in a non-deliberative way. Propagation is not under the learner's direct control as information is activated. Activation spreads because information has the structure it has in long-term memory. In contrast, learners can decide, based on utility they calculate according to a schema like AEIOU, whether to apply particular operations – learning tactics and strategies – to information currently active in working memory. Working memory is where the learner can exercise choice. Perceptual systems, built up over extensive experience, filter information from the external environment. That system and information in long-term memory are not systems available to controlled activation. For example, a learner may notice an instructional designer's cues such as italicized font, propositions in text having a particular format (e.g., "We define …"), and an option offered on a menu in a software application. Learners also may be ignorant of or overlook (not attend to) phrases and other instructive conventions an author intends to cue particular operations applied to particular information.

In this context, the learner exercises choice about operations, standards, the schedule of evaluations, and AEIOU accounts of learning activities that unfold in working memory. Examples related to the preceding external conditions the learner re-presents in working memory might be: a judgment that italics strongly predicts utility for highlighting the italicized text, choosing to postpone looking up confusing terms because an efficacy expectation forecasts later text can be analyzed to fill gaps of understanding, a reminder offered by the menu option *Tag*... signals it is possible to catalog (assemble) selected information in a way that eases locating it under future conditions, e.g., cramming for next week's exam.

Identifying the information underlying motivated operations can be a challenge for observers, especially when compactly unified patterns of behavior, cognition, metacognition, and motivation are bound together in automated, multi-event packages triggered and executed practically without the learner's awareness. An everyday example in my experience is making careful (rational, by my standards) word choices while enthusiastically promoting a controversial point to a friend while I'm in the midst of planning a turn at a traffic intersection crowded with cars, buses, and pedestrians.

In cases of motivated behavior we observers characterize as SRL, whether deliberative or automated, the IF-THEN-ELSE model begs for specifying what information constitutes IFS. As noted just above, motivation questionnaires do this in at least two ways: (a) describing a situational context within which to consider one's response to a generic experience or topic – this course, science; and (b) a particular state or experience – knowing one's own and others' scores on a measure of achievement. The question needing address in research carried out in dynamic online contexts is how to identify IFS learners identify in everyday learning activities that are gateways to THENS or ELSES.

3.2 Cognition

Instructional designs explicitly and implicitly guide learners about operations they might apply when working on tasks. Explicit directions may be provided by learning (instructional) objectives presented at the beginning of chapters and self-test questions appearing at the end of chapters. Implicit cues about selecting content on which to operate and tactics for learning can be observed, e.g., as headings for sections of chapters and "leading" questions embedded in text.

Such directions and cues have a 2-part grammar: task + topic. In this illustrative instructional objective, differently styled underlining marks task and topic: Develop an argument, pro or con, for reducing on-street parking to allow widening bike lanes. Arguments can be described by a schema with facets or slots such as: claim, evidence supporting the claim, and warrants validating evidence as appropriate to the claim. This basic argument schema can be expanded to include more than one instance of and multiple kinds of evidence. More complete arguments (a) add counterarguments shaped by this same schema but presenting the case opposite to the pro argument, then (b) end with a summary resolution balancing the pro and con presentations. The argument schema provides informational cues about kinds of information to search, how to assemble those information products when weighing costs and benefits of widening bike lanes that reduce on-street parking, and standards for evaluating a draft argument.

In many cases where an argument is assigned as an essay or in-class presentation, the learner engages three further tasks. A first is searching curated sources or the wide-open internet for information relevant to the proposition to be argued. A second is determining the credibility of evidence that will be selected and cited, a multi-operation process called sourcing (Braasch & Bråten, 2017). Sourcing involves evaluating properties of information in a source such as the author's credentials, characteristics of the medium of publication (e.g., blind reviewed publication vs. unmoderated posts in social media), and the presence and nature of boundary conditions the author provides for claims (e.g., Everyone knows … vs. In the case of one-way side streets …). The third major task is crafting the essay or talking points to form the argument per se.

Operationally defining data to record some operations when a learner engages in these tasks is straightforward. A learner's search for sources and information within them is easily logged when a learner enters words into a search engine or, after a source is loaded, a search box. Monitoring content for evidence can be traced if software provides tools for the learner to highlight text and tag those selections as *evidence*. Recording that a learner monitors properties of information regarding credibility can be tracked if tags are available to mark it as *trustworthy* vs. *doubtful*. Or, a structured note can cue monitoring these features by presenting a form with a text box labeled *evidence* followed by a checkbox list to monitor properties (standards) applied in evaluating the credibility of that evidence. Software features like these might be considered prompts or scaffolds designed to stimulate operations like monitoring and assembling. When learners use tools like these, individual or a package of operations can be traced because the learner operates on particular information.

3.3 Metacognition

When self-regulating learners track and adapt their engagements in learning, metacognition is applied in two ways. First, learners monitor information in working memory. That information is selectively imported from external sources and registered alongside information retrieved from long-term memory. This bundle of information can be monitored to classify its properties and rate its features. For example, a learner may judge a diagram is complicated, or a science lab experiment described on an assignment sheet is interesting. Products of these operations can activate additional information in long-term memory and supply standards for searching external sources for particular information.

Metacognitive monitoring is a relational concept involving two bins of information which Nelson and Narens (1990) labeled the object level and the meta level. In the preceding example of monitoring a diagram, the object level concerns information the diagram represents, e.g., the water cycle (e.g., see https://www.noaa.gov/education/resource-collections/freshwater/water-cycle; National Oceanic and Atmospheric Administration). The meta level refers to the learner's evaluation(s) of properties of that information. Is it complex vs. simple or unimportant? Is it clear or too complicated? Reaching a metacognitive judgment – e.g., the diagram is complex – is the product of monitoring not what the water cycle is – e.g., water changes states due to evaporation and condensation – or the meaning of terms like evaporation. Information monitored at the meta level concerns properties of object level information, e.g., the water cycle diagram has a degree of complexity, or certainty about the meaning of condensation is low. Tracking the learner's operations on information at the meta level might be inferred if an eye tracking system records relatively long focus on a particular area of interest in the diagram, suggesting effort; or if the learner enters condensation in a search tool. The information in focus or entered in the search box is the key to observing this metacognitive operation.

Operational definitions for metacognitive control include two sequential steps. First, monitoring information at the meta level generates a product in working memory. Second, a particular operation the learner controls is selected for execution because the product of that monitoring operation has particular properties. Metacognitive control thus has the form of an IF-THEN-ELSE event. The learner who monitors properties of the water cycle diagram and reaches a meta-level characterization that it is complex may next apply an assembling operation that analyzes the cycle as a step-by-step chain of sequentially paired states: rain falls on land, water runoff accumulates in a lake, lake water evaporates … etc. Software annotations where the learner can select from a numbered list to label each successive pair can trace this operation.

4 Integrating Information with Trace Data

Models proposed to describe cognitive, metacognitive, and motivational operations involve slots filled by information, the subject of an operation. Without information, there is nothing on which to operate. As learners self-regulate learning, they can monitor information describing properties and products of operations to decide how they will tailor next-chosen operations to satisfy motivation. Products can be results of operations on subject matter as well as results describing perceptions about operations, e.g., an operation's pace, effort required, and so forth. This leads to the proposition introduced earlier: Information is a necessary component when developing accounts of learning events modeled by IF-THEN-ELSE. How does this perspective apply to identifying and analyzing SRL?

4.1 Examining Effects of One Operation

Table 11.2 presents fabricated data for three learners' scores on four measures of achievement about chemical bonds. For each subject matter topic identified in a row of Table 11.2, software logged whether students applied or did not apply operation X to that topic. Columns on the right side of Table 11.2 record for each student their scores on some items gauging motivation, a test of knowledge or some metacognitive event relating to the topic. For example, data trace all three students applied operation X to subject matter information about the electron shell. Alex and Tracy indicated they were motivated to learn that topic (e.g., tagging it *interesting* or it merits effort to *review*), or learned it (e.g., correctly answered a practice quiz item) or metacognitively judged high confidence about it (e.g., typed the topic label into a note titled *Learned Concepts*). Kris' scores show the opposite.

Table 11.2 records identical total scores for each student. These were computed by summing item scores. Also shown is the conditional probability operation X generated an effect. This is computed by counting events where operation X is applied and the learner's score is 1, then dividing the sum of those "successful" events by the number of observed events. For Alex, on each occasion when operation X was applied, the score on a measure of whether the operation generated a "positive" product (positive motivation, achievement, positive metacognitive judgment) was 1. For learning events when Alex did not apply operation X, the product was not positive. In other words, operation X worked perfectly for Alex and any operation other than X was not productive (as gauged by a single measure of the product).

In Tracy's case, there is no discernable pattern relating using operation X and positive products.

Kris scored 1 on a product only if some operation other than X was applied. For Kris, operation X was consistently unproductive while some other operation was consistently productive.

All three students appear identically motivated, or equally cognitively or metacognitively engaged when their use of operation X is considered as an aggregate (total). But operations clearly had differential effects. Using aggregate scores, neither a learning scientist nor a learner receiving learning analytics to guide SRL could be clear about "what works," how operations relate to effects. Moreover,

Table 11.2 Data and conditional probability statistics measuring effects of operation X

		Score pattern		
Information	Operation X applied?	Alex	Tracy	Kris
Electron shell	Yes	1	1	0
Ionic bond	No	0	1	1
Covalent bond	Yes	1	0	0
Metallic bond	No	0	0	1
Total (sum)		2	2	2
Pr[effect | operation]		1.00	0.00	0.50

neither person can be alerted to opportunities to identify operations other than X that are consistently productive for learners like Kris. Nor would they be alerted to exploring IFs, conditions or evaluations, differentiating when operation X was productive for Tracy.

When data have patterns like those in Table 11.2, and when products of learning events are aggregated without identifying which operation was applied to which information, pinpointing the effects of an operation is indeterminate. Without fine-grained data about information operated on, decisions about updating an instructional design or a learner's decision policy guiding SRL can have erratic results.

4.2 How Information Enriches Trace Data About Operations

When learning events enacted by self-regulating learners are modeled in terms of IF-THEN-ELSE, operations implementing a learning tactic or strategy, THEN or ELSE, are initiated based on the results of a learner monitoring a bundle of conditions, the IFs. Fundamental IFs include:

- Internal information describing the learner's motivation cataloged by the AEIOU model.
- Knowledge the learner retrieves from long-term memory about the topic of the learning task.
- Features the learner perceives about the external learning context, e.g., access to supplementary content, help, tools available.
- Standards activated in working memory the learner will use to monitor properties of the learning event (e.g., pace, effort, confidence) and its product(s).
- Standards presented in the instructional design.
- Cues presented in the instructional design intended as guides for SRL.
- Information in sources, the subject to be learned.
- Information in learner-created artifacts – highlighted text, notes, etc. – representing products of the learner's operations on object-level (subject matter) information and on meta-level (properties of AEIOU, operations) information.

The last four entries in this list share an important and useful property. Each can be observed directly and with no or negligible intrusion on the learner's everyday approach to learning.

4.2.1 Operations Mark Conditions Learners Monitor

Content in sources learners study online can be delivered in a range of formats: words, symbolic expressions (e.g., mathematical relations, chemical reactions, graphic symbols), diagrams, graphs, photographs, animations, and more. Whatever the medium, self-regulating learners choose standards to monitor information at the object level – What does the information communicate about the subject matter being studied? – and at the meta level – What properties of mental state (e.g., motivation, frustration), operations (e.g., pace, effort), and object-level information (familiarity, complexity, clarity) characterize the current learning task? IF characteristics of information forming that bundle of conditions match the profile of standards currently in effect, THEN the learner exercises metacognitive control by applying a preferred operation. If not – ELSE – the learner self-regulates differently.

Operations learners enact can signal conditions have been monitored. This has a significant implication: Information in sources learners study and artifacts learners create as they study can be mined to identify standards self-regulating learners use to monitor IFs in learning events. For example, does a learner almost always select sentences defining constructs for highlighting? When a text refers to a diagram, does the learner scroll to display that diagram again or open a companion window to view the diagram alongside text describing it? When standards conveyed as information – italicized text, phrasing such as "As Fig. 5 shows …" – can be identified, a fuller picture of SRL can be painted by pairing those IFs with trace data reflecting operations, THENS. This coupling of conditions-as-information in sources with trace data sets a stage to develop conditional probability statements as illustrated in Table 11.2.

4.2.2 Standards Can Be Supplied Explicitly in Sources

Sources often plainly recommend standards learners might choose to monitor learning in the form of learning objectives. These cues explicitly name topics in a discipline, e.g., Newton's laws of motion or major products of a country; and kinds of information, e.g., principles and examples. Trace data describing SMART (or other) operations learners use is enriched by appending the topic(s) and kind(s) of information learners are cued to process.

Objectives also identify standards for tasks, e.g., define, apply, or analyze. Named tasks label schemas with slots for declarative information or steps in a structured procedure (script). For example, a *define* task might label a schema with slots: concept label, critical property 1, critical property 2 …, family membership, example. A procedural schema for graphing a straight line given a symbolic expression like $y = 3x + 5$ might proceed in steps: identify the intercept in the expression, plot the intercept point, identify the slope coefficient, starting at the plotted intercept move 1 x-unit to the right then upward if the coefficient is positive or downward if the coefficient is negative a number of units equal to the coefficient, plot the point, connect the two plotted points. Trace data reflecting operations learners apply as they create artifacts to accomplish a learning objective can be augmented by the subject matter information and task schema in the objective.

4.2.3 Information in Sources

When information in sources is formatted as text or can be automatically translated to text from other formats, such as videos or images, that information can be analyzed to identify concepts on which it would be predicted learners should operate as they learn. Several approaches are available.

Content creators often using conventions to format content as prompts for learners to operate on particular information. Examples include italicized and bolded words to prompt monitoring understanding, blue font in webpages to prompt a direct search for information to be assembled with information in a current source, arrow symbols in diagrams suggesting rehearsing a sequence or self-explaining why A → B, and numbered lists suggesting the learner activate an order-preserving mnemonic to store items. Learners' operations on formatted information can be traced. For example, consider a numbered list of sequenced steps describing a process. A 2-column note form – step/reason – can be designed to trace whether learners assemble an explanation describing how that process progresses from step to step. Re-listing steps in the note traces rehearsing of a step. If learners paraphrase the source, natural language processing (NLP) methods can gauge the semantic correspondence of each description to the source, indexing the operation of translating. A final text box in the note form labeled *Make a 1st-letter mnemonic* traces assembling information represented in steps as a unitized multi-step procedure.

Some sources learners study include a glossary. Its entries are subject matter concepts learners should engage as they study that source. Key concepts and related concepts can sometimes be automatically identified by cataloging HTML <a href>*link text*<a> tags. Phrasing conventions can be searched to identify key disciplinary concepts, e.g., "We define …" or "X is the [key, dominant, main …] factor in …." Keyword extraction algorithms also might be used to extract key concepts.

Terms in a source's text, in a provided glossary and terms learners create often are defined using other terms in the glossary. Based on this in-terms-of relation, software systems like nStudy (Winne et al., 2019) can relate terms via edges in a node-link graph, a termnet. Learners' artifacts – e.g., notes, selected and tagged text, described in the next section – can be analyzed using the termnet to identify whether they include terms and how learners assemble knowledge using those terms. A learner using terms in artifacts that the termnet relates directly signals rehearsing a meaningful assembly. When a learner's artifact includes terms, say A and D, related by traversing interventing nodes in the termnet, say A–B–C–D, this traces the learner assembling conceptual structures beyond those explicitly provided in the source's definitions. Walks across intervening nodes in a termnet graph suggest more about what a learner knows than just the text a learner enters in an artifact. As well, examining terms learners search relative to those included in their notes can traces gaps, represented as intervening terms in a termnet, the learner is searching because those gaps need filling to assemble a multi-node information structure.

4.2.4 Selections, Notes, and Tags

Learners commonly select text to highlight and as anchors for notes about subject matter (Miyatsu et al., 2018; Peverly & Wolf, 2019). Selections signal monitoring, and the text selected contains clues about why monitoring was executed. What standards does the learner use as governors for searching and monitoring which text to select?

Providing tags learners may choose to index content is expected to stimulate their search for content by standards the tags describe. Consider a learner studying a text about research methods in psychology. Providing tags such as *independent variable* and *confound* likely encourages the learner to activate standards for searching information about those types of variables. When selections are assembled with one of those tags, this is evidence of monitoring for particular kinds of variables. What the learner selects reveals information judged to be one or the other kind of variable.

In some software systems, like nStudy, notes can be designed by researchers or instructors to present schemas prompting learners' annotations. Slots in those schemas guide learners to assemble structured accounts of subject matter. Each schema can be labeled, e.g., ARGUMENT or EXPLAIN. Its slots, fields in which learners enter information, also can be labeled. When learners select (a) information to anchor a note and (b) a labeled note schema for the note they will make, this traces monitoring by the learner: the selected information has a role in the chosen schema. As the learner enters information in slots of the schema, the note artifact records which information the learner assembles according to that schema.

Beyond supplying more detailed data for analyzing conditional probabilities, illustrated by Table 11.2, notes could be leveraged by an algorithm to automatically generate self-test questions or self-explanations. For example, if the learner is annotating a step-by-step process with explanations, questions can be algorithmically constructed: "What process begins with [paste step 1]?" This question affords opportunity for the learner to monitor assembling the name of a process with its initiating step. Another question might be: "Why is it important that [paste step 2] precede [paste step 3]?" This prompts self-explanation, a learning event with proven value (Bisra et al., 2018). As well, such questions directly associate operations on information which the learner performed while studying with items measuring whether products of those operations match targets for achievement. As in Table 11.2, these data are more direct tests of effects operations have. As well, information for the learner to restudy can be recommended alongside learning analytics about which learning tactic was not successful in promoting achievement.

Selection artifacts, such as text or regions of a graph the learner highlights, can be counted as instances of metacognitive monitoring to gauge the learner's overall engagement. By examining what information learners select relative to structures like a termnet, models can be developed to describe the learner's attention to specific content. Coupled with the aforementioned automatically generated (self)test items, predictive models might be developed to gauge not just how much a learner is learning while they study, but also topics and kinds of content they can be prompted to process.

Information selections also provide meta-level information about rhetorical roles for the selected information, e.g., definitions, principles, examples, and so forth. Learners can be offered tags to classify selections by role, enriching traces of metacognitive monitoring by revealing the learner's attention to and use of metacognitive standards.

Tagging is already practiced by many learners. Perhaps the most widely used and most basic tag is the yellow (or blue or pink or …) highlight. It marks information selected by monitoring; selected information matches an unspecified standard that has utility for the learner. Tagging systems can operationally define those standards, making them observable. Some learners tag using symbols for selections. Examples are:? identifies information the learner metacognitively judges is vague or confusing.! marks especially important information. Modern software systems can offer multiple semantic and symbolized tags. Learners may be encouraged to use tags because tags can be applied to filter and retrieve selections, notes, and bookmarks tagged for particular purposes (e.g., nStudy; Winne et al., 2019). For example, a tag like *Huh?* could be used to filter all content about which a learner wants to seek help from a teaching assistant or peer. Follow-up data in the form of an online chat with peers or an email to the TA validates the learner's plan and subsequent execution.

Basic classes of tags might span four categories. Discipline-specific role tags mark information as an instance of a disciplinary class. In earth science, tags might classify information related to igneous, metamorphic, and sedimentary rocks. Rhetorical structure tags index content by roles information plays in a conceptual structure. These might include principle, example, and critical detail. Tags labeling tasks signal a learning event where selected information will be the subject of particular operations at some future time. Examples include: review, research, quotation (in an essay to be drafted). Affect tags can reflect a learner's monitoring of an emotional reaction to information. Instances might include: *wow!* (surprise), *duh* (boredom), and *cool* (interest). The information tags convey coupled with information tagged provides more precise tracing of SRL than simply counting instances of a monitoring event.

5 Analyzing Information-Enriched Trace Data

Almost all analyses of learning processes begin with data structured as a timeline of sequential events, often with timestamps marking onset or offset of the event. Some forms of analysis examine this data structure directly to identify patterns; e.g., an ABC pattern in x, m, k … ABC … x, y, z … ABC …. In some analyses, patterns allow for "skipping" intervening events bounded by a regular sequence of events initiating a pattern and another regular sequence terminating the pattern, e.g., an ABCDE pattern in x, m, k … ABCgDE … ABChDE … ABCjDE …. Others analyses transform the sequential timeline of events into a n × n matrix. This format records tallies for every possible pairwise sequence of events representing transitions from an initial event in a row to a follow-on event in a column. Every type of event (A, B, C …) in a transition can play the role of the initial event, condition in the COPES model, and a follow-on event, P in the COPES model.

Such "information-free" analyses of occurrence, frequency, timing, and patterning of operations ignore information learners operate on in learning events. Information is the condition that triggers any operation. And, information is the product of every operation. Omitting information from analyses of learning events classifies conditions and products as irrelevant to operations. As previously described, operations are "empty" in these analyses.

It is likely sophisticated extensions to conventional analyses of process data can be developed to incorporate information to which operations are applied. But relatively simple and straightforward analyses may suffice. Here is one example.

Suppose a learner is studying a unit about conic sections: circles, ellipses, hyperbolas, and parabolas. Sources the learner studies present terms (e.g., center, focus, major axis, eccentricity), equations describing each conic section and graphical examples of each. Among a variety of operations traced, consider two: translating and assembling. Classes of information rehearsed are terms (definitions) and examples. Examples can have two formats: text and graphs. In the source material the learner studies, there are:

- 8 terms (A, B, C, D, W, X, Y, Z), each with its definition
- 1 abstract equation for each conic section in which coefficients are variables (e.g., *a*, *b*)
- 1 example equation corresponding to each abstract equation in which coefficients are integers, and
- 1 graph of each conic section labeled with the integers appearing in each example equation.

The learner generates notes when studying this source:

- 4 notes: The definition of each term A, B, C, and D is copied (rehearsing) from the source and pasted in a note.
- 4 notes: The learner paraphrases (translating) the definition of each term W, X, Y, and Z.
- A note compares graphs of the parabola and hyperbola. The learner induces a principle (assembling), "As the coefficient of the vertex gets larger, the graphs extend farther from the origin."

If these 8 definitions are the only definitions in this source, the learner can be judged to have useful standards for monitoring information presented as a definition and is motivated to learn definitions. If the source contained, say, 20 definitions, there are several possibilities meriting analysis given the data in this example. This learner may have prior knowledge of the 12 (=20 – 8) definitions for which trace data were not generated. Or, the learner may lack clear standards for monitoring cues that mark a definition. This hypothesis could be tested in the next learning session by posting an instructional objective inviting the learner to tag definitions or, to leverage benefits of generative learning, create term notes. Roelle and Nückles' (2019) study suggests the latter guide for SRL will have differential effects depending on the source text's cohesiveness and density of elaborations, and whether the learner engages in retrieval practice. Cohesion can be gauged automatically using tools like Coh-metrix (McNamara et al., 2014). Retrieval practice can be promoted by an automatic question generation tool (e.g., see Das et al., 2021).

Suppose the learner recalls definitions A and D but not B and C. Rehearsing definitions appears not predictive of learning; odds are 1:1 applying the operation of rehearsing promotes learning. But an order effect – primacy, recency – may be operative if timestamps are considered.

Suppose the learner can recall definitions W, X, and Z but not Y. Translating (paraphrasing) definitions appears effective with odds 3:1, and translating definitions was more productive than rehearsing them. The order effect is moot when the learner translates definitions. A learning analytic based on these results could recommend the learner try to paraphrase definitions more often. As data accumulate across future learning sessions where subject matter changes, the potency and generalizability of translating definitions can be tested for $N =$ me. Future analytics can be refined as additional data accumulate.

Suppose data show, after the learner assembles a principle based on information in the source, graphing parabolas and hyperbolas given algebraic expressions is accurate. While slim, data support a conjecture: The learner understands how coefficients in algebraic expressions locate vertices for these conic sections. Odds cannot be proposed yet because there is only one instance of this conditional relation.

With big data sets describing each learner and homogenously formed clusters of learners displaying approximately equivalent learning signatures formed using information-rich trace data, this approach to analyzing data offers promise for guiding SRL at the same time helps advance learning science (Winne, 2022). The learner's SRL is depicted in ways that generate serviceable learning analytics. Moreover, variance in the learner's selections of operations invites investigating motivation and conditions that discriminate whether this learner uses particular operations to learn. The AEIOU model and theories on which it stands can guide that investigation, strengthening links between learning science and learning analytics.

6 Conclusion

Learners are ubiquitously self-regulating agents (Winne, 1995, 2018). In the context of an instructional design or the architecture of a website, learners select information targets they aim to learn and operations they will apply to learn. Information available in the environment and recalled to or generated in working memory is what learners think with and think about. Topics range widely: declarative and procedural knowledge comprising a discipline; metaknowledge about genres and presentation formats (text, tables, and graphics); fixed and emergent properties of tasks;

forecasts and feelings about learning tactics as steps to execute as well as perceptions about that execution across the lifespan of task engagement; and more.

This account leads to an important proposition: Processes – O in the COPES model and the SMART model elaborating operations learners apply in learning tasks – are insufficient to advance theory, research, and productive applications of learning science and learning analytics. To successfully model SRL as a process requires accounting for information in three ways implied by the IF-THEN-ELSE model of a learning event.

First, information, the IF, sets a stage for the learner to select subjects on which to operate and operations to apply. Conditions (C) in the COPES model of a learning task is a placeholder for the wide-ranging information a learner considers in relation to an about-to-be-executed operation, a THEN. Without data representing that information, the onset of learning processes is a mystery.

Second, as learners execute an operation, unless it is automated to the extent it proceeds without monitoring, properties describing that operation are generated. Some examples are pace, fluidity, and effort. These emergent properties are products learners can monitor relative to personal and externally recommended standards.

Third, beyond just noted products of an operation arising because a self-aware person executes the operation, operations also generate products transforming their subject, the curriculum. Monitoring these products relative to standards creates evaluations in two domains. One is the subject-matter per se, e.g., a summary of an article, a solution to a problem. The other is the bundle of motivations and emotions represented via incentives and attributions in the AEIOU model.

A great deal of data representing these kinds of information can be unobtrusively and almost immediately gathered when learners study online. Information can be analyzed when presented via text, figure and table captions, and images and speech automatically transcribed to text. Formatting via markup tags that deliver content provides data to detect properties of information. Labeled software and architectural features – e.g., labeled hyperlinks, labeled buttons (e.g., NEXT, BACK), search boxes where learners' queries can be recorded – unobtrusively deliver important data about information.

Other information internal to learners' thinking can be revealed by perceptively engineering traces. Ideal traces generate data across multiple elements of the COPES, AEIOU, and SMART models. For example, a learner making a note in the nStudy system selects text, chooses a particular schema for assembling information about that selection, and enters text and selections among options in labeled lists formatted as checkboxes, radio buttons, or a slider. Making notes is an everyday studying activity, a relatively unobtrusive technique to gather information about C, O, P, S, and potentially E depending on slots presented in the note's schema.

All this information should enrich accounts of learning events beyond records logging time-sequenced logs of "information-empty" processes. Because self-regulating learners regulate learning based on and generating information, merging this data gathered unobtrusively is a major step toward generating new and more useful theory for learning science. At the same time, by developing sharper accounts of the information learners can consider in SRL, learning analytics will be more strongly positioned to help self-regulating learners as learning scientists conducting their personal programs of research for N = me (Winne, 2022).

6.1 Next Steps

Incorporating information presented to learners and generated by them in studies of learning events can take some direction from basic characteristics of current instructional designs and build from sophisticated methods now coming into use.

First, subject matter disciplines are founded on and distinguished by, in part, key concepts of which they are constituted. Glossaries identify those concepts and afford a representation of the discipline's conceptual structure as a termnet constructed using the in-terms-of relation previously described. The field should improve on this representation to track and, when self-regulating learners request information or interventions introduce information for learners to consider, supply concepts for learners to consider based on conceptual structures fundamental to a discipline. A termnet offers one mechanism to do this.

Second, there is widespread acceptance and use of terminology describing tasks, perhaps most publicized in the form of the revised "taxonomy" cataloged by Bloom and colleagues (Anderson & Krathwohl, 2001; Bloom et al., 1956). These terms and their synonyms can be readily mined using NLP technologies applied to content learners study, including direct mention of tasks in learning objectives, and text they create as notes and essays. Blending termnets (or more sophisticated representations) with standards for judging these tasks provides resources for designing note schemas learners might use to assemble content, automatically generating (self) test items and monitoring content learners select for tagging and annotations. An especially intriguing possibility is to investigate the possibility of accurately predicting what a learner learns by analyzing trace data instead of having to administer a posttest following the study.

Third, process maps now generated to investigate how learners' operations are patterned (e.g., Saint et al., 2021) need extension. Information learners study instantiates a pattern that triggers operations learners apply based on their metacognitive knowledge about how to learn modeled by If-Then-Else. Analytical tools now used to examine patterns of process data empty of information need extension to incorporate the information units (e.g., schemas, rules) on which those processes operate.

This is an ambitious and exciting agenda. It merges state-of-the-art work in learning science, learning analytics, knowledge representation, NLP, and modeling of dynamic events. Big data about information learners study and tools they use to study are needed as raw material to fuel this research. Fortunately, that resource is becoming increasingly accessible as education and training migrates to online platforms supported by systems learners can use every day to study and complete assignments (see Winne, 2017).

References

Anderson, L. W., & Krathwohl, D. R. (2001). *A taxonomy for learning, teaching, and assessing: A revision of Bloom's taxonomy of educational objectives*. Longman.

Bandura, A. (1997). *Self-efficacy: The exercise of control*. W. H. Freeman. https://doi.org/10.1891/0889-8391.13.2.158

Bisra, K., Liu, Q., Nesbit, J. C., Salimi, F., & Winne, P. H. (2018). Inducing self-explanation: A meta-analysis. *Educational Psychology Review, 30*, 703–725. https://doi.org/10.1007/s10648-018-9434-x

Bloom, B. S., Engelhart, M. D., Furst, E. J., Hili, W. H., & Krathwohl, D. R. (1956). *Taxonomy of educational objectives: Handbook I: cognitive domain*. David McKay.

Braasch, J. L. G., & Bråten, I. (2017). The discrepancy-induced source comprehension (D-ISC) model: Basic assumptions and preliminary evidence. *Educational Psychologist, 52*, 167–181. https://doi.org/10.1080/00461520.2017.1323219

Chung, W. L., & Bidelman, G. M. (2022). Acoustic features of oral reading prosody and the relation with reading fluency and reading comprehension in Taiwanese children. *Journal of Speech, Language, and Hearing Research, 65*(1), 334–343. https://doi.org/10.1044/2021_JSLHR-21-00252

Das, B., Majumder, M., Phadikar, S., & Sekh, A. A. (2021). Automatic question generation and answer assessment: A survey. *Research and Practice in Technology Enhanced Learning, 16*(1), 1–15. https://doi.org/10.1186/s41039-021-00151-1

Dubovi, I. (2022). Cognitive and emotional engagement while learning with VR: The perspective of multimodal methodology. *Computers & Education, 183*, 104495. https://doi.org/10.1016/j.compedu.2022.104495

Flowers, P., Theopold, K., Langley, R., & Robinson, W. R. (2019). *Chemistry 2e*. OpenStax. https://openstax.org/details/books/chemistry-2e

Kuhn, T. S. (2012). *The structure of scientific revolutions*. University of Chicago Press. https://doi.org/10.1177/1745691617710510

McNamara, D. S., Graesser, A. C., McCarthy, P. M., & Cai, Z. (2014). *Automated evaluation of text and discourse with Coh-Metrix*. Cambridge University Press. https://doi.org/10.1017/CBO9780511894664

Miyatsu, T., Nguyen, K., & McDaniel, M. A. (2018). Five popular study strategies: Their pitfalls and optimal implementations. *Perspectives on Psychological Science, 13*(3), 390–407. https://doi.org/10.1177/1745691617710510

National Oceanic and Atmospheric Administration. (2019, February 01). *Water cycle*. https://www.noaa.gov/education/resource-collections/freshwater/water-cycle

Nelson, T. O., & Narens, L. (1990). Metamemory: A theoretical framework and new findings. In G. H. Bower (Ed.), *The psychology of learning and motivation* (Vol. 26, pp. 125–141). https://doi.org/10.1016/s0079-7421(08)60053-5

Peverly, S. T., & Wolf, A. D. (2019). Note-taking. In J. Dunlosky & K. A. Rawson (Eds.), *The Cambridge handbook of cognition and education* (pp. 320–355). Cambridge University Press. https://doi.org/10.1017/9781108235631.014

Renninger, K. A., & Hidi, S. E. (2019). *The Cambridge handbook of motivation and learning*. Cambridge University Press. https://doi.org/10.1017/9781316823279

Roelle, J., & Nückles, M. (2019). Generative learning versus retrieval practice in learning from text: The cohesion and elaboration of the text matters. *Journal of Educational Psychology, 111*(8), 1341–1361. https://doi.org/10.1037/edu0000345

Saint, J., Fan, Y., Singh, S., Gasevic, D., & Pardo, A. (2021, April). Using process mining to analyse self-regulated learning: A systematic analysis of four algorithms. In *LAK21: 11th international learning analytics and knowledge conference* (pp. 333–343). https://doi.org/10.1145/3448139.3448171

Vatansever, D., Menon, D. K., & Stamatakis, E. A. (2017). Default mode contributions to automated information processing. *Proceedings of the National Academy of Sciences, 114*(48), 12821–12826. https://doi.org/10.1073/pnas.1710521114

Weiner, B. (2010). The development of an attribution-based theory of motivation: A history of ideas. *Educational Psychologist, 45*, 28–36. https://doi.org/10.1080/00461520903433596

Winne, P. H. (1995). Self regulation is ubiquitous but its forms vary with knowledge. *Educational Psychologist, 30*, 223–228. https://doi.org/10.1207/s15326985ep3004_9

Winne, P. H. (2017). Leveraging big data to help each learner upgrade learning and accelerate learning science. *Teachers College Record, 119*(3), 1–24. https://doi.org/10.1177/016146811171190030

Winne, P. H. (2018). Cognition and metacognition within self-regulated learning. In D. Schunk & J. Greene (Eds.), *Handbook of self-regulation of learning and performance* (2nd ed., pp. 36–48). Routledge. https://doi.org/10.4324/9781315697048-3

Winne, P. H. (2020a). Construct and consequential validity for learning analytics based on trace data. *Computers in Human Behavior, 112*. https://doi.org/10.1016/j.chb.2020.106457

Winne, P. H. (2020b). A proposed remedy for grievances about self-report methodologies. *Frontline Learning Research, 8*, 165–174. https://doi.org/10.14786/flr.v8i3.625

Winne, P. H. (2022). Modeling self-regulated learning as learners doing learning science: How trace data and learning analytics help develop skills for self-regulated learning. *Metacognition and Learning, 17*, 773. https://doi.org/10.1007/s11409-022-09305-y

Winne, P. H. (in press). Assessing learning processes. In R. Tierney, F. Fizvi, K. Ercikan, & T. N. Hopfenbeck (Eds.), *International encyclopedia of education assessment and accountability* (4th ed.). Elsevier.

Winne, P. H., Gupta, L., & Nesbit, J. C. (1994). Exploring individual differences in studying strategies using graph theoretic statistics. *Alberta Journal of Educational Research, 40*, 177–193.

Winne, P. H., Teng, K., Chang, D., Lin, M. P.-C., Marzouk, Z., Nesbit, J. C., Patzak, A., Raković, M., Samadi, D., & Vytasek, J. (2019). nStudy: Software for learning analytics about processes for self-regulated learning. *Journal of Learning Analytics, 6*, 95–106. https://doi.org/10.18608/jla.2019.62.7

Chapter 12
Multimodal Measures Characterizing Collaborative Groups' Interaction and Engagement in Learning

Jonna Malmberg, Eetu Haataja, Tiina Törmänen, Hanna Järvenoja, Kateryna Zabolotna, and Sanna Järvelä

Abstract In this chapter, we outline how modes of interaction, such as cognitive and socio-emotional, and the regulation of learning provide support for collaborative engagement and examine how it changes over time. We start by framing how regulated learning is embedded in the cognitive and socio-emotional interaction between the group members from both a theoretical and a methodological perspective. We then move to illustrate, with an empirical case example, how multimodal data (i.e., video) and physiological signals, such as electrodermal activity indicating physiological synchrony between the group members, can be used to capture varying levels of collaborative engagement. The empirical example provides a complementary view on group interaction and collaborative engagement. We conclude by discussing how investigating group interaction that targets regulation can reveal how collaborative engagement is built and maintained. Additionally, we discuss future possibilities to harness multimodal data in practice to support collaborative engagement.

1 Introduction

Today, engagement is viewed as a multidimensional construct that involves behavioural, cognitive, and social forms, including self-regulated learning (SRL) (Cleary & Zimmerman, 2012; Fredricks et al., 2004). In the context of collaborative learning, these constructs are complementary to each other and are manifested through cognitive and socio-emotional interactions between collaborating learners. The concept of collaborative engagement builds on the self-regulated learning (SRL)

J. Malmberg (✉) · E. Haataja · T. Törmänen · H. Järvenoja · K. Zabolotna · S. Järvelä
Department of Educational Sciences and Teacher Education, Learning and Educational Technology Research Unit (LET), University of Oulu, Oulu, Finland
e-mail: jonna.malmberg@oulu.fi; eetu.haataja@oulu.fi; tiina.tormanen@oulu.fi; hanna.jarvenoja@oulu.fi; kateryna.zabolotna@oulu.fi; sanna.jarvela@oulu.fi

V. Kovanovic et al. (eds.), *Unobtrusive Observations of Learning in Digital Environments*, Advances in Analytics for Learning and Teaching,
https://doi.org/10.1007/978-3-031-30992-2_12

theoretical framework (Winne & Hadwin, 1998), because it allows to consider engagement as a dynamic process through which students participate in and maintain their engagement in collaboration over time (Cleary & Zimmerman, 2012). To conclude, extending the concept of collaborative engagement allows us to consider it as a process, and examine how its behavioural, cognitive and social facets change and build collaborative engagement over time, instead of considering it as an unchangeable inclusive state.

Socio-emotional and cognitive interaction involves collaborative and responsive interactions between group members (Isohätälä et al., 2017). Thus, we highlight how observable and covert forms of engagement are best understood within the social context, task and situation in which they occur (Cleary & Zimmerman, 2012). In collaborative learning, the key property of engagement is the interactional synchrony associated with the regulation of learning that occurs between individuals (Järvelä et al., 2016). A high degree of synchrony indicates a high level of collaborative engagement. In this chapter, we demonstrate how facets of engagement vary, including measures of physiological synchrony measured from the electrodermal activity (EDA) of the collaborating students. In empirical research, physiological synchrony has been shown to be informative and aligned with social interactions (Mønster et al., 2016) and aligned with the level of engagement (Hernandez et al., 2014; Khosa & Volet, 2014). Due to the rapid evolution of technology, educational research has begun to investigate new avenues to explore and augment theories of learning with novel technologies (Reimann & Bannert, 2017).

The large number of technological advancements in the field of education provides novel opportunities to explore collaborative engagement with unobtrusive multimodal data (Baker & Siemens, 2014) and consequently provides new viewpoints for collaborative learning and regulated learning research. Over the past few years, there has been an increasing interest in collecting multimodal data in the context of collaborative learning (Noroozi et al., 2020). Specifically, recent advances in combining multiple data channels (such as physiological data, log data, video recordings including gestures and utterances, and facial expressions) have made it possible to locate invisible markers of learner interactions, including regulated learning in the learning context (Malmberg et al., 2019a).

In the context of collaborative learning, the regulated learning process is a nuanced phenomenon that includes various representations in terms of cognitive and socio-emotional interactions (Järvelä et al., 2020). Contemporary research suggests gathering data from multiple sources can add to understanding of collaborative engagement and how it is shaped by regulation of learning (Azevedo, 2015; Lee et al., 2019). As well as the visible indicators, regulation of learning is influenced by physiological indicators such as stress, excitement, enthusiasm, or emotional dynamics (Mønster et al., 2016). Being able to capture these various multimodal representations in learning allows for richer understandings of how the learning process is regulated as it occurs. The power of multimodal data in SRL research lies in this capability to provide constitutive explanations that combine different modalities to unpack, for example, how sequences of different reactions and events change learners' regulated learning behaviour (Reimann, 2009).

In this chapter, we start by framing how regulated learning is embedded in the cognitive and socio-emotional interaction between group members from both theoretical and methodological perspectives. We then move to illustrate, with an empirical case example, how multimodal data, namely videos and electrodermal activity (EDA), can be used to capture varying levels of collaborative engagement. The empirical example provides complementary views on group interaction and collaborative engagement. We conclude by discussing how investigating cognitive and socio-emotional interactions, including regulated learning, can reveal how collaborative engagement is built and maintained within groups.

1.1 Engagement in Collaborative Learning

Collaborative learning is a complex combination of all learners' contributions to a groups' collective effort, reciprocal interactions, and joint attention (Barron, 2003). Learners in collaborative groups share information, search for joint solutions to the task and sustain a shared understanding of the task (Iiskala et al., 2011; Zabolotna et al., 2023). To engage in collaborative learning and achieve joint learning goals, learners must continuously monitor their learning and that of their group members' cognitive and socio-emotional interactions. Interaction among group members affects the quality and effectiveness of collaborative learning (e.g., Kuhn, 2015; Van den Bossche et al., 2006). Accordingly, collaborative learning requires learners' engagement and participation in joint activities towards a shared goal (Hadwin et al., 2018).

However, collaborative learning is seldom easy (Barron, 2003; Van den Bossche et al., 2006). Learners face socio-emotional or cognitive challenges that require them to recognize and externalize the challenges for their group members and engage in regulated learning (Hadwin et al., 2018). In the context of collaborative learning, socially shared regulation of learning (SSRL) and co-regulation of learning (CoRl) have been the main theoretical framework for understanding how students can overcome challenges in their engaged learning (Hadwin et al., 2018). Co-regulated learning (CoRL) occurs when learners' regulatory activities are guided, supported, shaped, or constrained by other members in the group (Hadwin & Oshige, 2011). Miller and Hadwin (2015) specified that in the context of collaborative learning, CoRL can take at least two forms. In the first form, CoRL occurs when group members prompt each other to contribute to the group. This happens, for example, when some group members prompt their peers to set learning goals that can be shared with the group. In the second form, CoRL occurs when an individual's SRL is gradually influenced by others. This means, for example, that other group members adapt their group members' learning goals but do not contribute to or co-construct learning goals together.

Socially shared regulation of learning (SSRL) emerges when group members work together to complement and negotiate shared perceptions and goals for the task. The group members then coordinate strategic enactments of the joint task, collectively monitor the group's progress and products, and make changes when needed to optimize collaboration in and across tasks (Hadwin & Oshige, 2011; Winne & Hadwin, 1998). In particular, the core of SSRL is the participation in transactive and reciprocal interactions, referring to the ways learners intentionally engage with and build upon each other's regulatory acts to solve cognitive or socio-emotional challenges in collaborative learning. This is to say, SSRL is embedded in cognitive and socio-emotional interactions. When learners engage in such interactions, they evaluate each other's contributions, justify and express their own opinions and ideas and provide answers to posed questions (Molenaar et al., 2011). This means that during the collaboration, students' interaction is not focused only on knowledge co-construction but also involves CoRL and SSRL embedded in cognitive and socio-emotional interactions that cannot be separated, especially in collaborative engagement (Järvelä et al., 2016).

1.2 Cognitive and Socio-Emotional Interaction Reflecting Students' Engagement in Collaborative Learning

Participation in interaction is the core mechanism facilitating collaborative engagement. It allows students to construct a shared conception of a problem (Roschelle & Teasley, 1995) and maintain the social and emotional conditions needed to keep the task progressing. In the collaborative learning literature, a distinction between cognitive and socio-emotional types of interactions is often made (Järvelä et al., 2016; Kreijns et al., 2003). In this chapter, we demonstrate the relevance of these two types of interaction for students for regulated learning and collaborative engagement.

Cognitive interactions refer broadly to a task-focused exchange among group members (Dillenbourg et al., 1995; Järvelä et al., 2016). Cognitive interaction is the engine building students' shared conceptions of a problem as it involves elaboration on the content to be studied (Sinha et al., 2015). In addition to discussing the task content, collaboration involves interaction that targets students' own and their group members' thinking. This means that students' interactions during collaboration are not solely focused on the task content itself but also involve metacognitive monitoring (Artz & Armour-Thomas, 1992). For example, it is important that students metacognitively monitor their progress and express their views in their interactions with other group members (Hadwin et al., 2018; Sinha et al., 2015; Malmberg et al., 2017). When the group members agree on how they understand the problem and how they are progressing with it, it is also easier for them to control and decide together on the efficient use of strategies for solving the task. This type of active interplay between monitoring and control has also been considered to reflect cognitive engagement in collaborative learning (Sinha et al., 2015). Possibly, due to that, the quality of monitoring seen in student interaction is linked to high-quality engagement. Students' monitoring of their own and their peers' task progress, task understanding, and task interests seem to contribute to group success. This means that high-quality monitoring asks for active and equal contributions to the process of group monitoring (Rogat & Linnenbrink-Garcia, 2011), supporting the socially shared regulation of the learning. However, recent research has shown that in addition to cognitive engagement, collaborative learning is, to a great extent, reliant on effective socio-emotional interaction.

In collaborative learning, group members can engage in a socio-emotional interaction as an operation to build up and maintain purposeful interchanges between students to express and shape perceptions of emotions and the group's socio-emotional atmosphere (Kreijns et al., 2003; Mänty et al., 2020). The quality of socio-emotional interactions indicates group members' collaborative engagement, which involves responsive and respectful interactions as well as group cohesion (Sinha et al., 2015). Previous research has found that positive socio-emotional interactions can promote cognitive engagement by facilitating CoRL and SSRL (Lajoie et al., 2015; Rogat & Adams-Wiggins, 2015; Rogat & Linnenbrink-Garcia, 2011). In moments of off-task behaviour, positive socio-emotional interaction can also be used as a means of supporting group members' behavioural engagement and to help them return to joint learning activities (Sinha et al., 2015; Järvelä et al., 2021). In turn, negative socio-emotional interactions can hinder the quality of group learning activities and, consequently, cognitive collaborative learning processes (Rogat & Adams-Wiggins, 2015). Negative socio-emotional interactions during group learning can also result in group members' negative emotional experiences of collaboration (Mänty et al., 2020) and, thus, play a role in how they will engage in future tasks. However, when negative socio-emotional interactions are challenging the group's grounds for collaboration, the group can utilize emotion regulation to restore a positive emotional state and to foster social engagement (Järvenoja et al., 2019). Accordingly, socio-emotional interactions and group regulatory processes, in combination, form a basis for understanding students' collaborative engagement as well as how group members collectively construct and maintain positive socio-emotional grounds for learning together (Järvenoja et al., 2013).

2 Studying Cognitive and Socio-Emotional Interactions as Part of Collaborative Engagement with Multimodal Data

Research on collaborative engagement is challenging as it stands at the intersection of individual and social processes. It is also challenging to show how types of interactions change over time in real-life learning settings and how learners regulate their learning to maintain engaging collaboration (Khosa & Volet, 2014). Over the past few years, a range of innovative analytical approaches for examining and interpreting the dynamics of interactions and regulated learning in real-life contexts have emerged (Azevedo, 2015). Emphasis in the field has been increasingly placed on real-time unobtrusive measurements that capture the dynamics of interaction and regulation as a part of engaged collaboration (Azevedo et al., 2018; Järvelä et al., 2016).

Recent research increasingly explores how multimodal data can be used to capture students' collaborative engagement and regulation of learning (Järvelä et al., 2020). This is because recent technological advancements have made it possible to utilize more data channels dealing with capturing cognitive and socio-emotional interactions within learning processes. Multimodal data is highly promising for collaborative learning research as it provides the potential to explain how self-regulation operates when learners engage with content. It can also extend our understanding on how collaborative engagement evolves over time in response to changes in situated conditions, and to present it as a function of changes in learners' level of engagement (Callan & Cleary, 2017).

Azevedo (2015) discusses how engagement can be detected by using unobtrusive multimodal data that capture cognitive, affective, metacognitive and motivational processes. Additionally, Azevedo (2015) evaluates how, and to what extent, different data channels (e.g., video data, log files, eye-tracking data and physiological sensors that capture EDA) are suitable for capturing engagement. Utilizing video recordings and physiological sensors is especially promising in the context of collaborative learning as it allows for the observation of participation in socio-emotional and cognitive interactions without interrupting students' learning process (Järvelä et al., 2019). This is because video data can provide continuous information about students' participation in such interactions that also coincides with, for example, study choices, confusion and changes in engagement in a learning situation, which are almost impossible to capture otherwise (Henriques et al., 2013; Winne, 2010). In its turn, EDA can provide continuous data related to perceived task difficulty and emotional activation (e.g., Malmberg et al., 2022a, b; Pecchinenda & Smith, 1996; Tomaka et al., 1993; Törmänen et al., 2022a). For example, Hernandez et al. (2014) investigated children's engagement in a social interaction with an adult by measuring EDA aligned with video data. They found that the features of EDA, such as the level of arousal measured from EDA and physiological synchrony, are relevant in detecting engagement during social interactions when compared with researchers' ratings of engagement during those interactions. Similarly, Morrison et al. (2020) found that measures of EDA were higher for items that the participating students rated themselves as being more engaged in the learning activities. This is to say, multimodal data, and especially the use of EDA in alignment with other data sources, has the potential to reveal collaborative engagement.

2.1 *Socio-Emotional Interaction Facilitates the Emergence of Group-Level Regulation*

Previous research has indicated that group members' collaborative engagement, defined by their participation in socio-emotional interactions, can facilitate the emergence of group-level regulation in collaborative learning (Lajoie et al., 2015; Rogat & Adams-Wiggins, 2015; Rogat & Linnenbrink-Garcia, 2011). Previous studies focusing on socio-emotional interactions utilizing multimodal data, such as video and EDA, have evidenced that when socio-emotional atmosphere remains generally positive, group members are more likely to initiate regulation in the face of socio-emotional challenges. However, when negative socio-emotional interactions are recurring, the groups' ability to regulate their learning is hindered (Törmänen et al., 2022a). Moreover, by utilizing EDA data as an indicator of the groups' emotional activation, their results propose that the group's shared regulation efforts and subsequent changes in their emotional states or learning activities may be reflected as changes in the physiological activation level. In addition, earlier research has shown that if individual students' socio-emotional profiles are different, it is likely to promote SSRL in a group (Törmänen et al., 2022a). Törmänen et al. (2022a) investigated individual group members' socio-emotional interaction profiles across four collaborative learning sessions with a person-centred approach. The results reveal three types of student profiles (negative, neutral, diverse), which can also be used as indicators of their social engagement in their group's collaborative learning. Students with a diverse profile are more likely to participate in their group's positive socio-emotional interactions than those with negative and neutral profiles, which can be considered an indicator of their high social engagement in collaborative learning. Accordingly, students with a diverse profile are more likely to contribute to the group-level regulation.

2.2 Cognitive Interaction Supports the Function of Group-Level Regulation

Previous studies have acknowledged the importance of SSRL for active engagement in collaborative learning interactions (Isohätälä et al., 2017). It has also been established that high-quality cognitive engagement depends on consistent metacognitive monitoring focusing on progress at the task (Sinha et al., 2015). Recently, Haataja et al. (2022) have studied how metacognitive interaction focusing on planning, task interpretation, strategy use, and reflection, group-level regulation and individual metacognitive monitoring accurately predict learning achievement in a high school physics course. Their results showed that the frequency of observed metacognitive monitoring that triggered CoRL was related to better learning achievement. However, the relationship of observed co-regulation to learning achievement depended on metacognitive monitoring that triggered cognitive interaction. This result emphasizes the importance of active collaborative engagement between group members because, besides potentially having an effect of its own, it could be the preacquisition needed for effective group-level regulation to occur.

In addition, previous studies have shown that when a group shows cognitive engagement in relation to challenging situations, they also tend to activate and align physiologically (Malmberg et al., 2021; Haataja et al., 2021). To specify, for example, Haataja et al. (2022) investigated how cognitive interactions, and more specifically interactions with a function of monitoring a group's collaborative learning process, relate to physiological arousal and physiological synchrony derived from EDA. The results show that, on average, groups' physiological arousal increased, and physiological synchrony was higher when groups monitored that they are not approaching their goal. To summarize, it seems that EDA has a great potential to inform the invisible and mental forms of need for regulation that provides support for collaborative engagement (Di Lascio et al., 2018).

2.3 Case Example – Analysis of Interactions in Engaged Collaboration

The case example illustrates what and how multimodal data (such as video recordings capturing students' collaborative interactions and EDA measuring their physiological synchrony) can indicate about students' collaborative engagement and regulation in collaborative learning.

2.3.1 Data Collection

The case example derives from the study design of secondary school students (~13 years of age, $N = 94$, 36 male, 58 female) from similar socio-economic backgrounds from a comprehensive school in an urban area of Northern Finland. The participating students were divided into 30 heterogeneous groups based on their previous science grade.

The study was conducted at the natural school settings as a part of their physics course. The students participated in the research when they were collaboratively learning about wave motion and its various physical manifestations during a 7-week study period. The collaborative tasks were designed together between the science teachers and researchers. The science teachers ensured that the topics covered the required subjects and contents, and the researchers ensured that tasks promoted regulation of learning and required genuine collaboration. The topics of the lessons and collaborative tasks focused on sound and light, light and vision, lenses, and reflection. For example, when studying reflection, the students were asked to use different types of lenses and make hypothesizes how the of beam of light would pass through different types of lenses and examine that in doing real experiments with the lenses. Each lesson followed the principles of the flipped classroom approach, due to its potential to facilitate the regulation of learning (Jovanović et al., 2019).

During their physics course the students were instructed to wear the Shimmer3 GSR (Burns et al., 2010) sensors independently to measure their EDA at the beginning of each physics lesson; they were informed that they could be taken off if they were uncomfortable. The lesson started with a short teacher-led instruction to ensure that students were familiar with the topic. This was followed by the collaborative learning tasks aimed to co-construct a more profound and shared understanding of the topic. The students' collaboration was also video recorded.

2.3.2 Analysis Protocol

The analysis proceeded in three phases. In phase 1, video data were coded to identify socio-emotional and cognitive interaction episodes. In phase 2, co-regulation and SSRL were identified from the coded socio-emotional and cognitive interaction episodes. In phase 3, physiological synchrony was observed from the EDA of each group member.

Phase 1. Locating Socio-Emotional and Cognitive Interactions from the Video Data First, *socio-emotional interaction* was coded in the 30-s segment when group members took verbal or behavioural actions related to socio-emotional or cognitive aspects of group formation and group dynamics, including expression of one's own emotions (Kreijns et al., 2003; Kwon et al., 2014). The code required interaction, which was defined as a reciprocal verbal exchange between two or more group members.

Socio-emotional interaction was coded when at least two group members showed clear verbal or visible bodily indicators of positive or negative emotions or engaged in negatively or positively charged interactions.

The coding scheme for cognitive interaction was developed and adjusted based on earlier coding scheme systems by Järvelä et al. (2016) and Whitebread et al. (2009). The first criterion to identify cognitive interaction was for students to engage in a task-focused interaction. The second criterion required at least two students to be involved in this interaction.

Phase 2. Locating Group-Level Regulation The second round of coding identified CoRL and SSRL from the coded socio-emotional and cognitive interaction episodes (Haataja et al., 2022). What differentiated these codes from socio-emotional and cognitive interaction was that students had to clearly express observation of an obstacle or a challenge in the learning process and, also, initiate regulation, which led to strategic changes in the groups' actions (Törmänen et al., 2022b). In CoRL, no additional response from other group members followed the initiation of regulation. In contrast, SSRL involved the reciprocal negotiated participation of several group members, leading to a strategic change in the learning process.

Phase 3. Measuring Physiological Synchrony Physiological synchrony reveals interdependence in physiology between the individuals in the group. The phasic signal component of EDA was used as the signal for calculation (Mendes, 2009). To calculate physiological synchrony, multidimensional recurrence quantification analysis (MdRQA) was used to quantify the physiological synchrony between the students.

3 Building Collaborative Engagement in Group Interaction – A Multimodal Data Case Example

Next, we present a case example that describes the first collaborative learning session from a group consisting of three female students (Linda, Maria and Rita). The case group was selected because it showed frequent cognitive and socio-emotional interactions, as well as mostly on-task behaviour. Further, in the first collaborative learning session, this group had frequent episodes of co- and socially shared regulation, which enabled the detailed exploration of these interaction processes in relation to each other. The case example aims to demonstrate the interplay between cognitive interaction, socio-emotional interaction, and regulation of learning, as well as their different functions in fostering group members' collaborative engagement. Further, the example uses physiological synchrony between the group members as a potential indicator of their collaborative engagement in the learning activity.

Figure 12.1 demonstrates the general flow of the physics lesson. During their collaboration, the group performed six subtasks (Task 1-Task 6) related to sound transmission which was the topic of the physics lesson. However, due to the nature of lesson structure, the teacher's instructions and organization of group work is not included in the description. As visualized in Fig. 12.1, the group showed cognitive interaction frequently throughout the session. They used cognitive interaction at the beginning of each subtask to form a shared task understanding, which they were also able to update while progressing with the tasks. Further, they used cognitive interaction frequently to metacognitively monitor their progress and reflect on their shared understanding of the phenomenon, which can be considered to signal collaborative engagement. In turn, the group used positive socio-emotional interaction, particularly during the first two subtasks, to build up a positive socio-emotional atmosphere for their collaboration, but also later to maintain a positive emotional state in the face of challenges and to promote the group members' social engagement. That is, the group showed high collaborative engagement throughout the learning process The case description shows how cognitive and socio-emotional interaction indicate students' collaborative engagement, but also set the stage for group-level regulation in the face of challenges.

Task 1: Mug Phone (0:39:00–0:50:00)

Building premises for collaborative engagement through cognitive and socio-emotional interaction.

0:39:00–0:43:30

The group starts a task on building and testing a mug phone. First, they start cognitive interaction on how to build the mug phone, which builds up cognitive engagement. After building the phone, they engage in positive socio-emotional interaction by laughing and having fun with the mug phone, which creates a positive emotional state for the group as a premise for their social engagement.

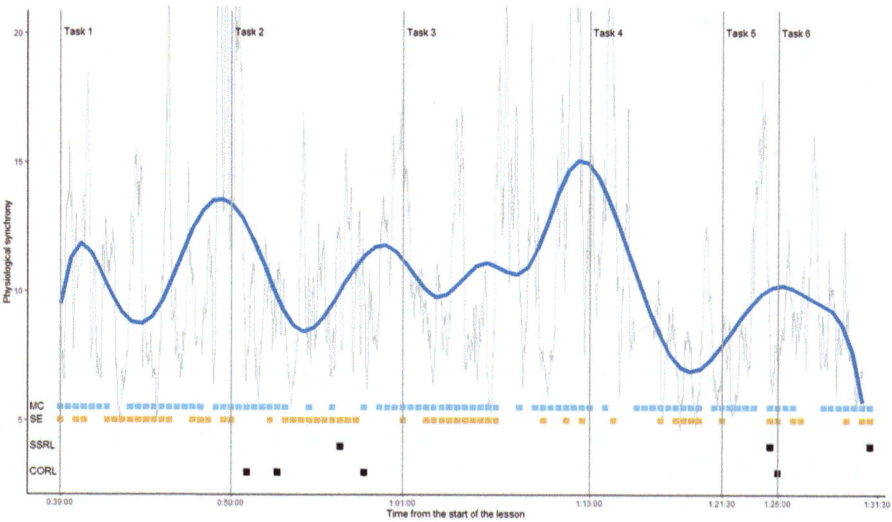

Fig. 12.1 Timeline of the case group demonstrating the occurrence of types of interactions and regulation along with physiological synchrony. Light blue marks cognitive interaction, orange marks socio-emotional interaction and black socially shared regulation and co-regulation. The blue line presents the trend of physiological synchrony of the group derived from the grey moving window MdRQA recurrence rate index

Moving towards the shared solution – physiological synchrony as a marker of collaborative engagement.
0:43:30–0:50:00

The group returns to the task instructions and, through cognitive interactions, builds a shared understanding of how they must explore the transmission of sound with the mug phone. Then, they start to execute the task together, discuss their findings, and agree on the answers they write down, showing a high cognitive engagement. Interestingly, while the group is moving towards the shared solution, the physiological synchrony between the group members starts to increase (0:45:00–0:49:30), potentially indicating the group members shared collaborative engagement in the learning activity.

Task 2: Church Bells (0:50:00–1:01:00)

Reorganizing for the new task – Decrease in physiological synchrony during individual activities.
0:50:00–0:51:00

The group starts a new task called "church bells," where they must explore the transmission of sound in a thread tied to a teaspoon. The group starts to prepare for the new task. Rita leaves the table to return the previous task equipment and the others start to foster social engagement through positive socio-emotional interaction towards the topic of the next task. Linda starts to joke ("Let's make the church bells!

I want to build the church bells. I have always dreamed of it!") and Maria joins ("Yay!"). Then, Linda and Maria start cognitive interaction and read the task out loud to form a shared task understanding. Rita returns to the table, but Linda and Maria leave to pick up the new task equipment. The group members perform different activities to prepare for the task, which seems to also be reflected in their physiological synchrony, which starts to decrease. Then, the group continues cognitive interaction together, aiming to form a shared understanding of the task by reading the instructions and discussing what they need to do in practice.

Engaged but not as a whole group.
0:51:00–0:53:30

Cognitive interaction in the group seems to prompt Linda to tell the others that she does not understand the task ("What do we need to do?") Maria responds by initiating co-regulation. She starts to tell Linda what she needs to do to perform the task. Linda and Maria start to explore the sound transmission together. However, Rita withdraws from the shared learning activity and starts to write down her notes individually. After a while, Linda and Maria face a cognitive challenge as they cannot hear the teaspoon through the thread. Linda uses positive socio-emotional interaction to maintain a positive emotional state in the face of the challenge and jokes that maybe they just have bad hearing. In turn, Maria starts cognitive interaction by considering the reasons why they cannot hear the sound ("What on earth? Why can't we hear the sound?"). Prompted by the metacognitive monitoring, Maria initiates co-regulation and suggests alternative strategies for the task execution, which Linda and Maria start to try.

Back in sync – having fun while learning as a whole group.
0:53:30–0:57:00

While still struggling with the task, Linda jokes again to maintain the positive emotional state ("We need to try all the pens available!") Finally, Linda and Maria succeed in the task and continue positive socio-emotional interaction, which supports Rita's behavioural engagement. Rita moves her attention back to the joint task execution and starts to make suggestions. When Rita returns to the joint activity, the physiological synchrony between the group members starts to increase again, along with the group members' collaborative engagement. To maintain Rita's behavioural engagement in the task, Linda and Maria ask Rita to try hearing the teaspoon and Linda continues positive socio-emotional interaction ("This is so fun, isn't it?") Then, the group continues the task execution together.

"This is so fun" – Maintaining collaborative engagement through socio-emotional interaction and regulation while reaching solution.
0:57:00–1:01:00

Linda and Maria start socially shared emotion regulation (Linda: "This is so, so fun!" Maria: "Yes, it is! We have so many observations related to this!"), promoting the group's social engagement in the task execution. The group continues task execution simultaneously, showing social engagement by having socio-emotional interaction on how the task is so fun. In addition to positive socio-emotional interaction, Linda uses co-regulation to ensure Rita's behavioural engagement with the task by asking if Rita has already written down some of their findings. Co-regulation prompts Rita to share her notes and the group starts cognitive interaction on how they can make sense of their findings and writes down their answers. The group reaches a solution and moves to the next task.

Task 3: Tuning Fork (1:01:00–1:13:00)

Moving to the next task – untuned again.
1:01:00–1:05:00

The group starts the third task: exploring the sound and wave motion with a tuning fork and water. First, they start cognitive interaction on task understanding. However, Maria and Rita leave to pick up the task equipment. Again, individual preparation activities seem to be reflected in a decrease in group members' physiological synchrony.

Increasing collaborative engagement through frequent cognitive and socio-emotional interaction.
1:05:00–1:13:00

The group has all the equipment ready, and they start enacting the task. They continuously engage in cognitive interaction to reflect their understanding of the phenomenon in light of their observations and findings showing high cognitive interaction. Further, they maintain social engagement through positive socio-emotional interactions (e.g., Linda: "This is so cool!") After the exploration, the group starts to discuss their shared answers. Along with the group members' increasing collaborative engagement in the task execution, their physiological synchrony seems to increase again while the group moves towards the task solution.

Task 4: Sound Volume (1:13:00–1:21:30)

Coordinating activity through cognitive interaction – No collaborative effort needed.
1:13:00–1:21:30

The group starts the next task: to categorize different sources of sound based on the volume. The nature of the learning activity changes from exploratory tasks to more traditional ones, where answers can be found in the textbook. In this task, the group members neither prepare together nor discuss how to proceed. Instead, Linda takes a lead and tells the others how she is going to do the task. The group follows Linda's lead, and they start to find the answers in their textbooks. The group stays coordinated in their learning activity through cognitive interaction and discusses the answers to form a shared understanding. Still, based on the decrease in group members' physiological synchrony, this task seems to be less optimal in fostering the collaborative engagement of the group. However, when the group decides to finalize the task, the physiological synchrony starts to increase again.

Task 5: How Deep Is the Lake? (1:21:30–1:25:00)

"Just calculations" – Cognitive interaction.
1:21:30–1:24:30

The group prepares shortly for the next task by cognitively interacting with and reading the task instructions out loud. The task is about calculating the depth of a lake based on the depth sounder information provided in the task instructions.

"We have certainly reached our goal" – Monitoring the progress with a positive tone.
1:24:30–1:25:00

After finding the correct answer for the task, Linda engages in cognitive interaction and monitors the group's progress by stating that they have almost performed all the tasks and they have only one task left. This initiates socially shared regulation in the group. First, Linda continues by monitoring that the group has certainly reached its goal. Rita then contributes by praising the group for completing all the tasks very thoroughly, simultaneously promoting collaborative engagement within the group. During this task, the group members' physiological synchrony increases again towards the end of the task.

Task 6: Transmission of Sound in Railways (1:25:00–1:31:30)

"…are you ready for the last task?" – Co-regulating collaborative engagement.
1:25:00–1:31:00

The last task starts when Linda promotes Maria's and Rita's behavioural engagement by co-regulation ("OK, are you ready for the last task?"). Linda is building up social engagement for the last task by initiating socio-emotional interaction by joking that she is an interviewer and starts to read the task out loud. This prompts, again, cognitive interaction within the group. The group starts to form a shared task understanding by reading the task together, and they also draw a picture of the calculation to increase their understanding. However, the teacher concludes the lesson, which interrupts the groups' task understanding phase.

Collaboratively engaged in reflection – building the foundations for next lesson.
1:31:00–1:31:30

Finally, after finishing their collaboration, the group shows cognitive engagement and starts socially shared regulation to reflect their goal achievement. Meanwhile, they maintain social engagement with positive socio-emotional interaction on how the lesson was so fun. Reflecting both, cognitive and socio-emotional aspects with a positive tone may set a fruitful foundation for the group's future lessons.

4 Practical Implications and Future Potential of the Research Reviewed

With the case example, we demonstrated the strengths and weaknesses that different data channels hold for characterizing interaction and engagement in collaborative learning. Regarding collaborative engagement and the different types of interactions that constitute it, i.e., cognitive and socio-emotional, video data offers invaluable evidence of the occurrence of these types of interactions that show how students engage in collaboration. Physiological data complements the observations and offers an affirmation that reduced collaborative engagement of a group is also reflected in decreased physiological synchrony. This means that when students are working individually with the task, physiological synchrony decreases. In contrast, highly engaging episodes seem to co-occur with cognitive and socio-emotional interaction visible in the video, in addition to increase in physiological synchrony.

A growing body of empirical research has demonstrated that when a group has frequent cognitive interactions throughout their learning process, it has the potential to support the function of group-level regulation as well (Haataja et al., 2022; Khosa & Volet, 2014). In turn, when group members participate in socio-emotional interactions, they are more likely to contribute to their group's regulation of learning (Törmänen et al., 2022a). Moreover, frequent positive socio-emotional interaction, in general, has the potential to foster the emergence of group-level regulation in the face of challenges (Bakhtiar et al., 2018; Törmänen et al., 2022b). The findings of these earlier empirical studies show the function of cognitive and socio-emotional interaction for group-level regulation but do not, however, reveal how they intertwine and are realized in actual collaborative interaction. The detailed case example made visible how individual contributions for regulation, cognitive- and socio-emotional interactions promote each other, exist in parallel and function equally, without any subordinate relationship in either direction. Yet, we still lack systematic empirical research showing how cognitive- and socio-emotional interaction both foster group-level regulation that provides support for collaborative engagement. The case examples illustrate how collaborative engagement is built and maintained temporally and is guided by the situational conditions. This is to say, engagement is not an enduring state, but is rather shaped in collaborative interaction.

Video data is a valuable source for understanding the qualities of interaction. What we can see and hear provides contextual information on how collaborative engagement is built up in a learning situation. However, video analysis is highly time-consuming and labour-intensive, even when systematic approaches are applied (Malmberg et al., 2019a, b; Zabolotna et al., 2023). Nevertheless, video data is still needed to fully understand the situated nature of collaborative engagement (Järvenoja et al., 2015). It makes concrete and visible the moments of physiological synchrony, when all group members contribute to task execution through cognitive and socio-emotional interactions and reveals how collaborative engagement is manifested in a situation. One way, perhaps, to speed up the laborious video-analysis could be to explore the potential of speech recognition (e.g., frequency of individual contributions) aligned with physiological synchrony (Noroozi et al., 2019). Combining speech recognition and data resulting from EDA with traditional video observations could potentially reveal the moments of collaborative engagement, but this requires further examination.

Yet, the question arises of whether and to what extent EDA can be used to measure collaborative engagement. Since EDA reflects the general level of physiological arousal, it remains a debatable question whether such data and i.e., indices of physiological synchrony on their own can offer much information about the state of collaborative engagement and regulation. This is to say, EDA should not be treated as the final authoritative data source for studying engagement, but rather used as a complementary method. Combining it with traditional video-based analyses is a good example of how learning processes can be examined from different perspectives by using multiple data sources. There may be a potential for future artificial intelligence (AI) technologies to help expedite such work. The reliability and appropriate use of AI-based technologies will also depend on what and how generic data are used to develop them and to what extent they can be applied in varying learning tasks. However, the current chapter provides an interesting viewpoint on the possible ways to explore collaborative engagement with the existing unobtrusive methods.

Acknowledgements Blinded.

References

Artz, A. F., & Armour-Thomas, E. (1992). Development of a cognitive-metacognitive framework for protocol analysis of mathematical problem solving in small groups. *Cognition and Instruction, 9*(2), 137–175. https://doi.org/10.1207/s1532690xci0902_3

Azevedo, R. (2015). Defining and measuring engagement and learning in science: Conceptual, theoretical, methodological, and analytical issues. *Educational Psychologist, 50*(1), 84–94. https://doi.org/10.1080/00461520.2015.1004069

Azevedo, R., Taub, M., & Mudrick, N. V. (2018). Understanding and reasoning about real-time cognitive, affective, and metacognitive processes to foster self-regulation with advanced learning technologies. In D. H. Schunk & J. A. Greene (Eds.), *Handbook of self-regulation of learning and performance* (2nd ed., pp. 254–270). Routledge. https://doi.org/10.4324/9781315697048

Baker, R., & Siemens, G. (2014). Educational data mining and learning analytics. In R. K. Sawyer (Ed.), *The Cambridge handbook of the learning sciences* (2nd ed., pp. 253–272). Cambridge University Press. https://doi.org/10.1017/CBO9781139519526.016

Bakhtiar, A., Webster, E. A., & Hadwin, A. F. (2018). Regulation and socio-emotional interactions in a positive and a negative group climate. *Metacognition and Learning, 13*(2), 57–90. https://doi.org/10.1007/s11409-017-9178-x

Barron, B. (2003). When smart groups fail. *Journal of the Learning Sciences, 12*(3), 307–359. https://doi.org/10.1207/S15327809JLS1203_1

Burns, A., Doheny, E. P., Greene, B. R., Foran, T., Leahy, D., O'Donovan, K., & McGrath, M. J. (2010). SHIMMER: An extensible platform for physiological signal capture. In *2010 annual international conference of the IEEE Engineering in Medicine and Biology Society (EMBC)*, pp. 3759–3762. https://doi.org/10.1109/IEMBS.2010.5627535

Callan, G. L., & Cleary, T. J. (2017). Multidimensional assessment of self-regulated learning with middle school math students. *School Psychology Quarterly, 33*(1), 103–111. https://doi.org/10.1037/spq0000198

Molenaar, I., Chiu, M. M., Sleegers, P., & van Boxtel, C. (2011). Scaffolding of small groups' metacognitive activities with an avatar. *International Journal of Computer-supported Collaborative Learning, 6*, 601–624.

Cleary, T. J., & Zimmerman, B. J. (2012). A cyclical self-regulatory account of student engagement: Theoretical foundations and applications. In S. L. Christenson, A. L. Reschly, & C. Wylie (Eds.), *Handbook of research on student engagement* (pp. 237–257). Springer. https://doi.org/10.1007/978-1-4614-2018-7_11

Di Lascio, E., Gashi, S., & Santini, S. (2018). Unobtrusive assessment of students' emotional engagement during lectures using electrodermal activity sensors. *Proceedings of the ACM on Interactive, Mobile, Wearable and Ubiquitous Technologies, 2*(3), 1–21.

Dillenbourg, P., Baker, M. J., Blaye, A., & O'Malley, C. (1995). The evolution of research on collaborative learning. In E. Spada & P. Reimann (Eds.), *Learning in humans and machine: Towards an interdisciplinary learning science* (pp. 189–211). Elsevier.

Fredricks, J. A., Blumenfeld, P. C., & Paris, A. H. (2004). School engagement: Potential of the concept, state of the evidence. *Review of Educational Research, 74*(1), 59–109. https://doi.org/10.3102/00346543074001059

Haataja, E., Malmberg, J., Dindar, M., & Järvelä, S. (2021). The pivotal role of monitoring for collaborative problem solving seen in interaction, performance, and interpersonal physiology. *Metacognition and Learning, 17*(1), 241–268. https://doi.org/10.1007/s11409-021-09279-3

Haataja, E., Dindar, M., Malmberg, J., & Järvelä, S. (2022). Individuals in a group: Metacognitive and regulatory predictors of learning achievement in collaborative learning. *Learning and Individual Differences, 96*, 102146. https://doi.org/10.1016/j.lindif.2022.102146

Hadwin, A. F., Järvelä, S., & Miller, M. (2018). Self-regulation, co-regulation, and shared regulation in collaborative learning environments. In D. H. Schunk & J. A. Greene (Eds.), *Handbook of self-regulation of learning and performance* (pp. 83–106). Routledge.

Hadwin, A., & Oshige, M. (2011). Self-regulation, coregulation, and socially shared regulation: Exploring perspectives of social in selfregulated learning theory. *Teachers College Record, 113*(2), 240–264.

Henriques, R., Paiva, A., & Antunes, C. (2013). On the need of new methods to mine electrodermal activity in emotion-centered studies. In L. Cao, Y. Zeng, A. L. Symeonidis, V. I. Gorodetsky, P. S. Yu, & M. P. Singh (Eds.), *ADMI 2012: Agents and data mining interaction* (pp. 203–215). Springer. https://doi.org/10.1007/978-3-642-36288-0_18

Hernandez, J., Riobo, I., Rozga, A., Abowd, G. D., & Picard, R. W. (2014). Using electrodermal activity to recognize ease of management in children during social interactions. In *UbiComp'14: Proceedings of the 2014 ACM international joint conference on pervasive and ubiquitous computing*, pp. 307–317. https://doi.org/10.1145/2632048.2636065

Iiskala, T., Vauras, M., Lehtinen, E., & Salonen, P. (2011). Socially shared metacognition of dyads of pupils in collaborative mathematical problem-solving processes. *Learning and Instruction, 21*(3), 379–393. https://doi.org/10.1016/j.learninstruc.2010.05.002

Isohätälä, J., Järvenoja, H., & Järvelä, S. (2017). Socially shared regulation of learning and participation in social interaction in collaborative learning. *International Journal of Educational Research, 81*, 11–24. https://doi.org/10.1016/j.ijer.2016.10.006

Järvelä, S., Järvenoja, H., Malmberg, J., Isohätälä, J., & Sobocinski, M. (2016). How do types of interaction and phases of self-regulated learning set a stage for collaborative engagement? *Learning and Instruction, 43*, 39–51. https://doi.org/10.1016/j.learninstruc.2016.01.005

Järvelä, S., Gašević, D., Seppänen, T., Pechenizkiy, M., & Kirschner, P. A. (2020). Bridging learning sciences, machine learning and affective computing for understanding cognition and affect in collaborative learning. *British Journal of Educational Technology, 51*(6), 2391–2406.

Järvelä, S., Järvenoja, H., & Malmberg, J. (2019). Capturing the dynamic and cyclical nature of regulation: Methodological Progress in understanding socially shared regulation in learning. *International Iournal of Computer-supported Collaborative Learning*, 425–441.

Järvelä, S., Malmberg, J., Haataja, E., Sobocinski, M., & Kirschner, P. A. (2021). What multimodal data can tell us about the students' regulation of their learning process?. *Learning and Instruction, 72*, 101203.
Järvenoja, H., Järvelä, S., & Malmberg, J. (2015). Understanding regulated learning in situative and contextual frameworks. *Educational Psychologist, 50*(3), 204–219.
Järvenoja, H., Näykki, P., & Törmänen, T. (2019). Emotional regulation in collaborative learning: When do higher education students activate group level regulation in the face of challenges?. *Studies in Higher Education, 44*(10), 1747–1757.
Järvenoja, H., Volet, S., & Järvelä, S. (2013). Regulation of emotions in socially challenging learning situations: An instrument to measure the adaptive and social nature of the regulation process. *Educational Psychology, 33*(1), 31–58.
Jovanović, J., Mirriahi, N., Gašević, D., Dawson, S., & Pardo, A. (2019). Predictive power of regularity of pre-class activities in a flipped classroom. *Computers & Education, 134*, 156–168. https://doi.org/10.1016/j.compedu.2019.02.011
Khosa, D. K., & Volet, S. E. (2014). Productive group engagement in cognitive activity and metacognitive regulation during collaborative learning: Can it explain differences in students' conceptual understanding?. *Metacognition and Learning, 9*, 287–307. https://doi.org/10.1007/s11409-014-9117-z
Kreijns, K., Kirschner, P. A., & Jochems, W. (2003). Identifying the pitfalls for social interaction in computer-supported collaborative learning environments: A review of the research. *Computers in Human Behavior, 19*(3), 335–353. https://doi.org/10.1016/S0747-5632(02)00057-2
Kuhn, D. (2015). Thinking together and alone. *Educational Researcher, 44*(1), 46–53. https://doi.org/10.3102/0013189X15569530
Kwon, K., Kim, E. M., & Sheridan, S. M. (2014). The role of beliefs about the importance of social skills in elementary children's social behaviors and school attitudes. *Child & Youth Care Forum, 43*, 455–467. https://doi.org/10.1007/s10566-014-9247-0
Lajoie, S. P., Lee, L., Poitras, E., Bassiri, M., Kazemitabar, M., Cruz-Panesso, I., Hmelo-Silver, C., Wiseman, J., Chan, L. K., & Lu, J. (2015). The role of regulation in medical student learning in small groups: Regulating oneself and others' learning and emotions. *Computers in Human Behavior, 52*, 601–616. https://doi.org/10.1016/j.chb.2014.11.073
Lee, V. R., Fischback, L., & Cain, R. (2019). A wearables-based approach to detect and identify momentary engagement in afterschool Makerspace programs. *Contemporary Educational Psychology, 59*, 101789. https://doi.org/10.1016/j.cedpsych.2019.101789
Malmberg, J., Järvelä, S., & Järvenoja, H. (2017). Capturing temporal and sequential patterns of self-, co-, and socially shared regulation in the context of collaborative learning. *Contemporary Educational Psychology, 49*, 160–174.
Malmberg, J., Järvelä, S., Holappa, J., Haataja, E., Huang, X., & Siipo, A. (2019a). Going beyond what is visible: What multichannel data can reveal about interaction in the context of collaborative learning?. *Computers in Human Behavior, 96*, 235–245.
Malmberg, J., Haataja, E., Seppänen, T., & Järvelä, S. (2019b). Are we together or not? The temporal interplay of monitoring, physiological arousal and physiological synchrony during a collaborative exam. *International Journal of Computer-Supported Collaborative Learning, 14*, 467–490.
Malmberg, J., Fincham, O., Pijeira-Díaz, H. J., Järvelä, S., & Gašević, D. (2021). Revealing the hidden structure of physiological states during metacognitive monitoring in collaborative learning. *Journal of Computer Assisted Learning, 37*(3), 861–874.
Malmberg, J., Saqr, M., Järvenoja, H., & Järvelä, S. (2022a). How the monitoring events of individual students are associated with phases of regulation: A network analysis approach. *Journal of Learning Analytics, 9*(1), 77–92.
Malmberg, J., Haataja, E., & Järvelä, S. (2022b). Exploring the connection between task difficulty, task perceptions, physiological arousal and learning outcomes in collaborative learning situations. *Metacognition and Learning, 17*(3), 793–811.
Mänty, K., Järvenoja, H., & Törmänen, T. (2020). Socio-emotional interaction in collaborative learning: Combining individual emotional experiences and group-level emotion regula-

tion. *International Journal of Educational Research, 102*, 101589. https://doi.org/10.1016/j.ijer.2020.101589

Mendes, W. B. (2009). Assessing autonomic nervous system activity. In E. Harmon-Jones & J. S. Beer (Eds.), *Methods in social neuroscience* (pp. 118–147). Guilford.

Miller, M., & Hadwin, A. (2015). Scripting and awareness tools for regulating collaborative learning: Changing the landscape of support in CSCL. *Computers in Human Behavior, 52*, 573–588.

Mønster, D., Håkonsson, D. D., Eskildsen, J. K., & Wallot, S. (2016). Physiological evidence of interpersonal dynamics in a cooperative production task. *Physiology & Behavior, 156*, 24–34. https://doi.org/10.1016/j.physbeh.2016.01.004

Morrison, A. L., Rozak, S., Gold, A. U., & Kay, J. E. (2020). Quantifying student engagement in learning about climate change using galvanic hand sensors in a controlled educational setting. *Climatic Change, 159*, 17–36. https://doi.org/10.1007/s10584-019-02576-6

Noroozi, O., Alikhani, I., Järvelä, S., Kirschner, P., Juuso, I., & Seppänen, T. (2019). Multimodal data to design visual learning analytics for understanding regulation of learning. *Computers in Human Behavior, 100*, 298–304. https://doi.org/10.1016/j.chb.2018.12.019

Noroozi, O., Pijeira-Díaz, H., Sobocinski, M., Dindar, M., Järvelä, S., & Kirschner, P. A. (2020). Multimodal data indicators for capturing cognitive, motivational, and emotional learning processes: A systematic literature review. *Education and Information Technologies, 25*, 5499–5547. https://doi.org/10.1007/s10639-020-10229-w

Pecchinenda, A., & Smith, C. A. (1996). The affective significance of skin conductance activity during a difficult problem-solving task. *Cognition and Emotion, 10*(5), 481–503. https://doi.org/10.1080/026999396380123

Reimann, P. (2009). Time is precious: Variable-and event-centred approaches to process analysis in CSCL research. *International Journal of Computer-Supported Collaborative Learning, 4*, 239–257.

Reimann, P., & Bannert, M. (2017). Self-regulation of learning and performance in computer-supported collaborative learning environments. In *Handbook of self-regulation of learning and performance* (pp. 285–303). Routledge

Rogat, T. K., & Adams-Wiggins, K. R. (2015). Interrelation between regulatory and socio-emotional processes within collaborative groups characterized by facilitative and directive other-regulation. *Computers in Human Behavior, 52*, 589–600. https://doi.org/10.1016/j.chb.2015.01.026

Rogat, T. K., & Linnenbrink-Garcia, L. (2011). Socially shared regulation in collaborative groups: An analysis of the interplay between quality of social regulation and group processes. *Cognition and Instruction, 29*(4), 375–415. https://doi.org/10.1080/07370008.2011.607930

Roschelle, J., & Teasley, S. D. (1995). The construction of shared knowledge in collaborative problem solving. In C. O'Malley (Ed.), *Computer supported collaborative learning* (pp. 69–97). Springer. https://doi.org/10.1007/978-3-642-85098-1_5

Sinha, S., Rogat, T. K., Adams-Wiggins, K. R., & Hmelo-Silver, C. E. (2015). Collaborative group engagement in a computer-supported inquiry learning environment. *International Journal of Computer-Supported Collaborative Learning, 10*(3), 273–307. https://doi.org/10.1007/s11412-015-9218-y

Tomaka, J., Blascovich, J., Kelsey, R. M., & Leitten, C. L. (1993). Subjective, physiological, and behavioral effects of threat and challenge appraisal. *Journal of Personality and Social Psychology, 65*(2), 248–260. https://doi.org/10.1037/0022-3514.65.2.248

Törmänen, T., Järvenoja, H., Saqr, M., Malmberg, J., & Järvelä, S. (2022a). A person-centered approach to study students' socio-emotional interaction profiles and regulation of collaborative learning. *Frontiers in Education, 7*, 866612. https://doi.org/10.3389/feduc.2022.866612

Törmänen, T., Järvenoja, H., Saqr, M., Malmberg, J., & Järvelä, S. (2022b). Affective states and regulation of learning during socio-emotional interactions in secondary school collaborative groups. *British Journal of Educational Psychology*, e12525. https://doi.org/10.1111/bjep.12525

Van den Bossche, P., Gijselaers, W. H., Segers, M., & Kirschner, P. A. (2006). Social and cognitive factors driving teamwork in collaborative learning environments: Team learning beliefs and behaviors. *Small Group Research, 37*(5), 490–521.

Whitebread, D., Coltman, P., Jameson, H., & Lander, R. (2009). Play, cognition and self-regulation: What exactly are children learning when they learn through play? *Educational and Child Psychology, 26*(2), 40–52.

Winne, P. H. (2010). Bootstrapping learner's self-regulated learning. *Psychological Test and Assessment Modeling, 52*(4), 472.

Winne, P. H., & Hadwin, A. F. (1998). Studying as self-regulated learning. In D. J. Hacker, J. Dunlosky, & A. C. Graesser (Eds.), *Metacognition in educational theory and practice* (pp. 277–304). Erlbaum.

Zabolotna, K., Malmberg, J., & Järvenoja, H. (2023). Examining the interplay of knowledge construction and group-level regulation in a computer-supported collaborative learning physics task. *Computers in Human Behavior, 138*, 107494. https://doi.org/10.1016/j.chb.2022.107494

Chapter 13
Electrodermal Activity Wearables and Wearable Cameras as Unobtrusive Observation Devices in Makerspaces

Victor R. Lee

Abstract Makerspaces are a unique type of environment for unobtrusive observation of learning in digital environments given that they encourage free movement and open interactions with a range of tools and people. Wearable devices that can generate and collect data about the wearer's skin conductivity offer some new opportunities for conducting research on the learning that takes place in makerspaces. This chapter summarizes three studies and their analysis approaches that have used the combination of wearable electrodermal activity devices and wearable cameras to identify moments suggestive of high levels of youth engagement. The specific considerations that went into designing data analysis procedures for this environment are discussed as are the eventual solutions that were deployed in these studies. While use of wearable electrodermal activity is still new for in situ maker education research, the early results suggest that it may contribute to our understanding of what triggers engagement. Namely, free and active social interaction is identified as an especially important quality to preserve in makerspaces and maker-oriented learning experiences. While inferences like this and others could be made, this chapter also firmly asserts that still more work remains to be done to help the field settle on best analytical practices for using these wearables and analyzing the resultant data to maximize their potential for unobtrusively observing and analyzing engagement in these complex and dynamic environments.

Keywords Electrodermal activity · Wearable devices · Wearable cameras · Skin conductance · Makerspaces · Maker education · Maker movement · Youth engagement

V. R. Lee (✉)
Stanford University, Stanford, CA, USA
e-mail: vrlee@stanford.edu

V. Kovanovic et al. (eds.), *Unobtrusive Observations of Learning in Digital Environments*, Advances in Analytics for Learning and Teaching,
https://doi.org/10.1007/978-3-031-30992-2_13

1 Introduction

We are now into the second decade of the "Maker Movement" in education (Martin, 2015). This means that it is not uncommon to find makerspaces in public and private primary and secondary schools, although they may go by other names such as fabrication labs (FabLabs), STEAM Labs, design studios, or some combination of those words. Public libraries increasingly provide makerspaces and maker programs among their community offerings (Melo & Nichols, 2020). Universities, especially those with engineering programs and schools, now often have multiple makerspaces on their campuses and linked with their courses. Dedicated university courses exist related to making and maker pedagogy (e.g., Fields & Lee, 2016). The general characterization of "making" promoted through these is that in an era of mass production and consumption, it is now increasingly possible to make things that would have historically been only possible in large companies with high-end professional machining and fabrication equipment (Dougherty, 2013). At the same time, there is also some acknowledgment that making is a core part of the human experience and robust across cultures and communities (Vossoughi et al., 2016). Making may be most frequently associated with dedicated engineering spaces, 3D printers, and circuitry, but it has long existed in heritage craftwork, family practices, and hobbies.

The educational interest is based on a number of beliefs, some of which have theoretical backing and some that are, at least at present, based more on intuition. One belief is that making is an especially effective way to enhance learning. By actively working with physical and digital materials, we are able to discover, test, and explore new ideas that can be embodied and enacted with materials (Papert, 1980). Some of this may be increased disciplinary content knowledge – for example, working with electronic textiles can be a means for learning more about circuitry (e.g., Peppler & Glosson, 2013). Other boosts to learning, conceived as growth in knowledge, may take the form of richer, more actionable knowledge that comes with doing actual projects that may reach across disciplinary boundaries or be novel solutions to problems (Blikstein, 2013).

Another belief is that making tends to challenge typical educational structures and routines. For instance, who is deemed knowledgeable may change, and an overall culture of mentorship and peer support may be engendered in making (Sheridan et al., 2014). Kids using makerspaces take on engineering approaches and see multiple solution possibilities for problems. Being positioned as creators and authors may also engender a "growth" mindset (Dweck, 2006) as youth see their abilities increase over time.

And one last belief, which is the focus of this book chapter, is that something about the experience of making or learning in makerspaces really supports youth engagement (e.g., Dougherty, 2013). Makerspaces could be a place where young people can really explore their own interests given autonomy and a number of materials. It could be more lively and enjoyable for youth than the arrangements that are common in a typical classroom with desks facing toward the front. Perhaps even just being a creator is more motivating or more satisfying for a large segment of children because of the pride one gets from completing the work and having a finished product at the end.

For a handful of years, a strand of my research has sought to explore the speculations, conjectures, and hypotheses that circulate about making and youth engagement. It was in many ways an extension of work I had been long pursuing related to the potential of wearable devices in education (Lee, 2019; Lee & Shapiro, 2019). At the onset of this work, wearable devices were becoming available that could capture the experience of the wearer and enable data capture without being tethered to a desktop computer. Most notably, wrist-worn devices that could detect and record electrodermal activity (EDA) – also known as skin conductance level, the amount of electrical conductivity facilitated by the skin – were being produced and were available for purchase by researchers. This was a major shift from the typical skin conductance measurement apparatuses used in psychophysiological research. Those more traditional setups would often involve sitting at a computer workstation with a wired device attached to one's hand or finger, sometimes with conductive jelly, in a very strictly climate-controlled room. The wearer needed to remain stationery given the wiring. However, wearable EDA sensing made it far more feasible for non-specialists to do skin conductance measurement and allow for a research participant's free, untethered movement in space. With EDA wearables, we could conduct unobtrusive observations of learning in digital environments – and for makerspaces, the digital environment was not necessarily confined to a screen; it was the entire space that had, among other things, digital fabrication tools.

This chapter summarizes some of the work that we have done with EDA wearable devices coupled with wearable cameras, which are also instruments for making unobtrusive observations. This work had been made possible through funding from the National Science Foundation in the United States (Grants **CNS-1623401** and **CNS-1949740**) and enabled us to really invest research time to ascertain how much we could see in terms of youth engagement in makerspaces with wearable instrumentation.

2 Related Literature

The literature on making as a paradigm for learning environments and learning experiences has benefitted from several years of research and scholarly argument. More is still being produced. However, the curious reader is encouraged to refer to some synthesis volumes, such as the *Makeology* books (Peppler et al., 2016a, b) that brought together many different authors and researchers of maker education. For this chapter, the literatures to be discussed are those related to engagement and related to the use of wearable EDA sensing for educational research.

2.1 Engagement

"Engagement" is a term that is often used to refer to an aspiration or state of being for students, but it is not always treated with precision. Fredricks et al. (2004) provide one synthesis of how educators have treated and implicitly defined the topic of engagement. Decades prior, it was largely understood as participation and commitment to schooling and was operationalized as amount of school attendance. This treated engagement as being the same as "commitment". As engagement became a more intentional focus in education and educational psychology research, seeing students completing specific activities and exhibiting signs of focus and willingness to do related work came to be understood as "engagement" but at a finer-grained level than whether or not someone had been simply present at their school a given number of days. Fredricks et al. offered in their synthesis a characterization of engagement as a construct with cognitive, behavioral, and affective dimensions and rather fluid time scales and relationships to other existing constructs, such as motivation, strategy use, or interest. The treatment I have pursued speaks more to the situation where we observe something that looks like focused commitment to a specific task on the scale of a few seconds of time. It may be productively and more precisely understood as "situational engagement" that is cued in the moment and can be described through the three dimensions mentioned above (cognitive, behavioral, and affective). The cognitive dimension would speak both to the intent use of cognitive resources – the moments of more intense or greater amounts of mental investment whether it be in the use of working memory, cuing of prior knowledge, or involvement of motivational beliefs or goals that support the active and intentional use of cognitive resources. The behavioral dimension would speak to direct observables of a student in action. For instance, we may notice that when a student is sitting on the edge of their seat and leaning forward, we are inclined to call them engaged. Or, we may see them in a span of several seconds enacting movements and actions that are highly consistent with what are deemed necessary for the deliberate and purposeful completion of a specific task (e.g., when ascertaining if a student is behaviorally engaged in mathematical computation, they are writing with pencil and paper and referring frequently to a nearby calculator and textbook featuring a word problem rather than aiming a paper airplane at their friend sitting halfway across the room). Finally, the affective dimension relates to how one feels with respect to emotion, valence, and disposition in those seconds of researcher or educator interest. Often, we think of engagement positively and in terms of enjoyment. However, one could be very engaged while administering CPR after an automobile accident or shedding tears of sadness in response to a tragic scene in a film. However, yawning (because the person feels boredom) when a friend is sharing a personally important opinion would suggest lack of affective engagement. Together, these three dimensions are taken as better describing engagement at the situational, momentary (i.e., a few seconds) time scale.

Building upon this tripartite treatment of engagement and scale of observable learning activities (on the scale of seconds to minutes), Sinatra et al. (2015) offered some additional ways to approach the study of engagement as part of a journal special issue on youth engagement in science. In addition to remaining cognizant of and affirming the three dimensions identified by Fredricks et al., Sinatra et al. presented a methods-oriented continuum that varied from person-oriented, person-in-context, and context-oriented approach for conceiving of, documenting, and analyzing engagement. A person-oriented treatment would often study a single individual in a controlled laboratory setting and rely on measures such as reaction time, self-report, or eye movement (e.g., Miller, 2015). On the other end, context-oriented approaches would tend to be observational research done in situ with deliberate attention to what aspects or features of the context produce highly engaged behavior at a given moment during a larger activity, and presumably also highly engaged cognition and positively valenced affect. Renninger and Bachrach (2015) were offered as an example of this as it was an observational study of environmental triggers for situational engagement – an event that could be the starting point for development of interest (Hidi & Renninger, 2006).

Person-in-context approaches would be somewhere in between. The question would be how specific individuals are experiencing a complex learning experience that may not be under the full control or design of the researcher. Experience sampling is one potential method for doing such work (e.g., Xie et al., 2019) as it would rely on random alerts or prompts for a research participant to log in the moment of interruption what was their current state and circumstances (supporting the seconds to minutes time scale). While it can be effective at generating rich and abundant in situ data, it does require attentional resources and interruption. A more unobtrusive approach to capturing engagement from a person-in-context could involve wearable devices that are continuously operating without obvious interruptions to the wearer. It is what my research group and I pursued, albeit erring more toward seconds rather than minutes for instances of situational engagement.

2.2 *Wearable Electrodermal Activity Sensing*

Above, I had summarized how a classic paradigm for obtaining electrodermal activity data had been with an apparatus wired to a computer or workstation in a controlled environment. The interest in this line of psychophysiological research is based on some now-accepted findings regarding changes in skin conductivity in response to various conditions (see Dawson et al., 2007 for a textbook-level treatment much of which is abbreviated and summarized here). Skin includes sweat glands that are activated as an unconscious physiological response by the sympathetic nervous system. Hands and feet are known to have a large number of sweat glands relative to many other parts of the body. In response to stimuli, the sweat glands increase in their activity – along with several other sympathetic nervous system activities such as heart rate and pupil dilation. It turns out that even before we produce enough sweat to form a visible bead or drop, the conductive properties of our skin change as the sweat glands activate. This takes place in a short window of time that varies between 0.5 and 5 s, at least as ascertained by laboratory studies when stimuli known to trigger a sympathetic nervous system response (e.g., a large image of a spider) are presented. Plotted against time, the skin conductance level looks like a "peak" (Fig. 13.1). More precise terminologies such as tonic and phasic components of the measured conductivity are used, but the combined rapid ascendence and descendance in conductivity are the key artifacts of interest and we find easily conveyed to others as "peaks". While this response appears to be quite common, it is important to note it is not universal. There are individuals who do not exhibit it (such as people with Schizophrenia, Gruzelier & Venables, 1972) or just generally have low overall skin conductivity. However, it has been associated in other research studies with increased cognitive demand (Setz et al., 2010) and with cued state anxiety (Carrillo et al., 2001; Naveteur et al., 1987).

EDA has been noted as having compelling potential for person-centered engagement research (Azevedo, 2015). Relatively recently, in research by Poh et al. (2010), evidence was obtained suggesting that not only the hands and feet are effective areas to measure and discern skin conductivity patterns (and peaks), but so was the part of the forearm near the wrist. This subsequently led to the creation and marketing of wrist-based wearables that looked comparable to a wristwatch or activity tracking band that cleverly had conductive nodes positioned near the wrist. Making the equipment wearable and wireless enables person-in-context research on engagement. One device that supported this type of methodological inquiry was the *Affectiva* Q Sensor. It had been used in studies of young children's engagement (Hernandez et al., 2014) and developed in prototype software for teachers to detect engagement levels in their classrooms (Daily et al., 2015). Another device that came out later and served to replace the discontinued Q sensor was the *Empatica* E4 device. Some studies comparing the E4 to traditional EDA instrumentation found medium to high correlations with traditional, wired-to-the-computer EDA instruments (Milstein & Gordon, 2020).

In the past several years, E4 devices had been used to detect and measure engagement in university lectures (Di Lascio et al., 2018), K-12 classrooms (Zhang et al., 2021), augmented reality environments (Soltis et al., 2020), digital coding

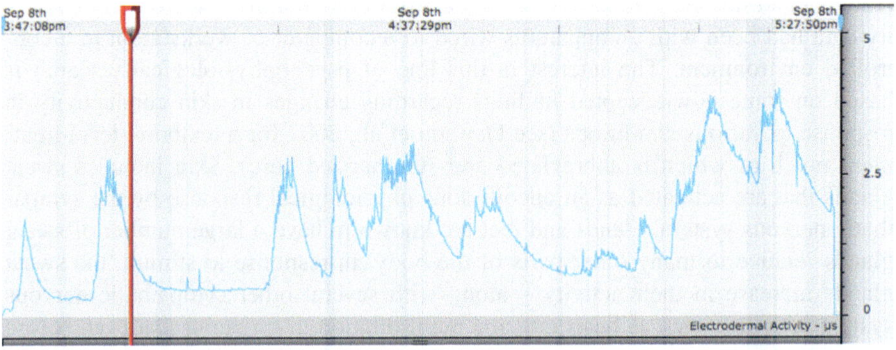

Fig. 13.1 EDA profile over time with peaks and troughs in skin conductivity readings

environments (Lal et al., 2021), youth theater programs (Eisenhauer, 2019), and even at conferences (Gashi et al., 2019). As one of many multimodal inputs, it has begun to appear in research related to making (Worsley & Blikstein, 2018). While its use is broadening to a range of settings, there are many unanswered questions about the use of wearable EDA data for observing learning. Gold standard methods and thresholds for recognizing EDA peaks when the environment is not carefully controlled have yet to be identified, and there are studies that raise questions about the range of populations and tasks for which EDA is useful (Betancourt et al., 2017) beyond those listed above, for making the same inferences that would typically be made with data from wired EDA devices (Menghini et al., 2019). More remains to be understood, but given some years of working with these devices, I can share some findings on the use of wearable EDA devices – both the Q sensor and the E4 – to capture situational youth engagement in makerspaces.

3 Empirical Person-in-Context Research with EDA

As our team was working on this larger project designing and conducting highly exploratory research with new instrumentation in complex settings, there were several considerations and strategic decisions involved in our research. The standards for how things should be done in these circumstances do not exist. Ultimately, we acted on based on our joint best judgment given the following:

- With respect to EDA, the makerspace environment would produce very noisy data. Recall that classic skin conductance research and accepted findings had been done with different instrumentation (wired systems connected to computers) and in highly controlled circumstances. These controls included climate control, which can go as far as controlling humidity as well as temperature as those could influence sweat gland activity. Research designs often took the form of controlled experiment with clearly defined individual stimuli. This would involve standardized tasks and displays. In contrast, makerspaces are lauded for their fluid activities and for the ability of a young learner to move freely based on their individual interests within them. This means that control over stimuli was relinquished. Environmental and climate conditions could not be fully controlled without great difficulty.
- Furthermore, our in situ approach meant that the participant samples were what they were based on who was available and had already agreed to be part of the activity in the makerspace. This meant that how many people showed up, at what time, and for how long were outside of our control. Because these were young people who were brought by caregivers for a multitude of reasons for finite periods, prescreening was very difficult. All participants and all those who were potentially documented in the research record gave their informed consent and assent. However, the number of people involved may have been below ideals for

statistical power for broader population generalizability, simply because there were no other people participating at the makerspace.
- The technologies used were still early releases of products that may have been sufficient for commercial distribution but still buggy and with limited usability. Not all devices purchased worked properly. Some malfunctioned. Some were accidentally disabled by the youth wearing them because of counter-intuitive interfaces and high potential for erroneous button-pressing. This is to be expected and accounted for, but in studies when the sample had a ceiling, these losses mean that some data that are collected will have limited use because of incompleteness or accidental user corruption of the data.

Given these challenges that were embedded in this research, one could question whether this investigation would even be worth pursuing. My contention is that it is so long as the analysis and recommendations are presented in ways that reflect deliberate caution and care. Even if the data obtained end up supporting a claim that eventually becomes refuted, it will still do the work that research is supposed to do. It builds ideas and practices that can be subject to further examination, refinement, and perhaps replacement. Transparency and caution are important to maintain. In light of that, our commitments for EDA data were the following:

- Current practice within psychophysiological research is to treat "peaks" in data as the sympathetic response and indicative of increased arousal. Thus, relative increases were to be foregrounded rather than absolute values. For example, if over the course of an hour, skin conductance levels were to constantly increase at a constant rate, we would see absolute increase in conductance but no "peaks" where the rate of change abruptly increased. Since peaks have been accepted in EDA research (as tonics and phasics), we opted to stick with peaks as the signal of interest. Otherwise, the inference from a linear increase in conductance levels would be that whatever took place at time i was suggestive of more sympathetic arousal than whatever took place at time i-1, and there were too many obvious alternative explanations (increased body heat and subsequent sweat activation, for example).
- Not *all* peaks should be treated as consequential because of variation and noise. Because the EDA data were expected to be noisy, there would be many small increases and decreases in EDA values because of both natural variation and error. Thus, we could only accept some of the increases as being "peaks" of note. To be conservative, we sought to only focus on those peaks that were on the higher end of the distribution. However, each person is different. The amount of skin conductance for two different people should not be thought of as the same. (Indeed, we had some youth who always had very low EDA readings even when we tried using multiple and different devices and even some small experiments to trigger peaks, such as suddenly intentionally making unexpected noises to startle. These individuals had EDA readings that never exceeded 1 microsiemen. We treated those as 'nonresponsive' and eliminated them from further EDA analyses.) This meant that we should not only look for the top segment of the distributions of "peaks" but we should also consider those peaks relative to each

individual. For those who remained after removal for never exceeding the 1 microsiemen threshold, some could have EDA data that would look more labile and some that looked more constant. As we looked at visualizations of youth peaks aggregated, we saw roughly normal distributions. Therefore, our strategy was to only include those relative increases in EDA activity that were greater than one standard deviation above the mean of all relative changes for that youth on that day's readings. At a minimum, it eliminated a majority of peaks. Using normal distributional assumptions, this would only include something like the upper 16% of relative changes in EDA levels.
- Finally, in an effort to be conservative in inclusion, we should actively remove as many peaks from the remaining subset if given reasons based on the literature and in light of our constraints. To accomplish this, we followed up directly with the authors of Taylor et al. (2015) who had published a system that was trained using machine learning to emulate expert EDA analysts to identify artifacts that should be excluded from an EDA data set. We applied their screening program to our data and our already-reduced set of peaks (from the bullet above, by only retaining those that were at least a standard deviation or more of an increase) and for any segments that the trained system identified as potentially problematic artifacts, removed them from our already reduced-set of notable peaks.

With these as considerations and commitments, we ended with a reduced set of timestamps when a youth exhibited a "peak" in their EDA data. We interpreted those as candidate moments for when there had been moments of abrupt, increased arousal. With those timestamps in hand, we then retroactively reviewed the first-person camera footage to get a record of what was being encountered, as could be discerned from a chest-worn camera. (We note that some important activity and where a youth was looking may not be represented in the camera footage.) As the EDA "peak" response takes place from 0.5 to 5 s prior to the appearance of a peak in highly controlled settings, what precisely in the environment may have triggered that peak would be inherently difficult to discern. Moreover, we assume that something in the environment acted as a trigger. It could have been possible a private thought or something internal to the youth's body that would not be observable for an outsider triggered the peak. As such, we operate under the assumptions that we can identify candidate experiences and qualities from the video record that would plausibly trigger the EDA response.

All of the above serves to demonstrate that this entire project was, at its core, principled guesswork. However, I would maintain that is the core of academic social science research. We build upon foundations from prior work and see how far we can get by relying on the same assumptions with new extensions. The broader social and historical endeavor will ultimately determine what has staying power and what does not, just as much as theories and paradigms eventually shift as our methods and arguments advance further (Kuhn, 1962).

With the above caveats and disclosures in mind, I now turn to summarizing some of our efforts in context and what we believe we have found. Much of this involves inventing methods for each study and sharing what inferences and informed speculations followed given the use of those different methods. While our analysis approaches varied, peaks were used as a part of the analyses in all three studies.

4 Study 1: EDA and Wearable Still Image Cameras in a Maker Project

The first study to which I was attached involving wearable EDA to examine engagement in a makerspace followed two girls, ages 10 and 12 (referred to by the pseudonyms "Dot" and "Jane" respectively), who were partnered together for a large scale youth maker club project at a community makerspace that involved launching a weather balloon with a sensor payload to obtain data from the atmosphere (Cain & Lee, 2020). This was initiated by the founder of the makerspace and head of the camp program and involved using the Ardusat space and atmospheric science DIY sensor program using Arduino controllers (the lessons and program now reside at becauselearning.com). A dozen participants were involved and worked in pairs with individual sensor input devices to install and prepare for inclusion with the weather balloon launch. The program took place over 12 weeks with a weekly 2-h meeting and work time at the local makerspace with supervision and support from makerspace staff and volunteers.

Three sessions in the makerspace were recorded. Dot and Jane each wore *Affectiva* Q Sensors on their non-dominant wrist. They also wore a specialized "lifelogging" camera from a company called Autographer from a hanging strap around their necks. The Autographer camera was designed for people who wished to obtain a record of their daily activities, especially as a potential memory prosthesis such as for individuals with memory challenges or conditions such as Alzheimer's disease. This device automatically took timestamped pictures based on a proprietary algorithm between every 8 and 15 s depending on detected changes in movement and lighting conditions (see Fig. 13.2). During a full day, the Autographer could capture and store approximately 2000 images. For this study, Dot and Jane only wore and used the Autographer for the 2-h sessions they were in the makerspace.

The Q sensors were configured to make 4 skin conductance recordings per second. Across the two girls and 3 days of makerspace activity as part of this makerspace project, they produced about 57,600 EDA readings and 4500 photographs. In addition to these wearable devices, we had standing video cameras to obtain a record of what was happening in the makerspace and focused on Dot and Jane as they worked in the makerspace.

Following systematic coding of still images obtained by the Autographer aligned with EDA peaks following the above-summarized procedures, we noted that while Dot and Jane were ostensibly working together as a pair on the same project, the recorded distribution of peak responses by activity differed. We computed "arousal ratios" to enable comparison by which the moments coded as specific types of makerspace activities, such as "watching adult model a task" or "soldering" and labeled

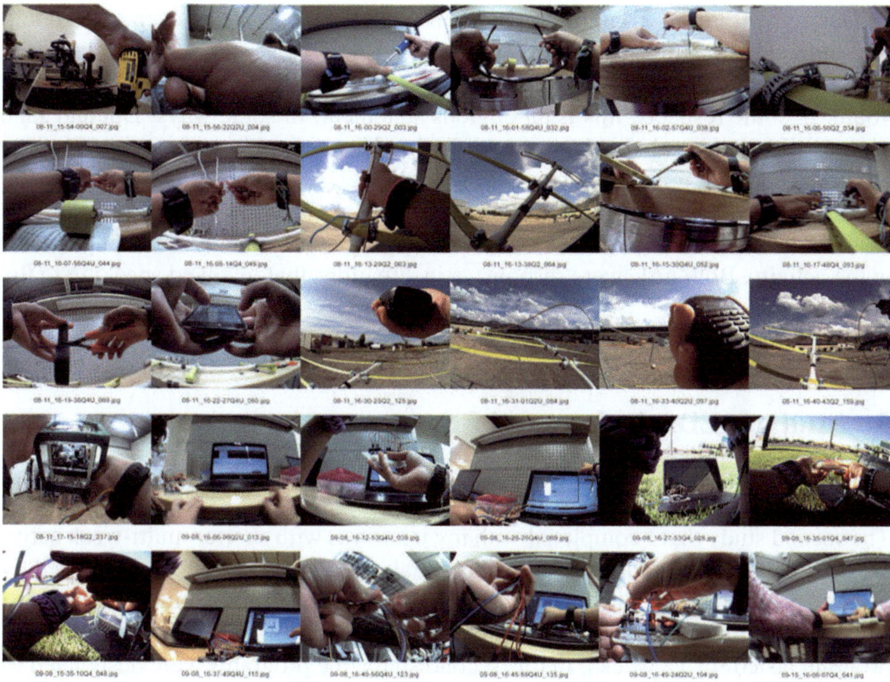

Fig. 13.2 Series of images captured by Autographer wearable camera when worn by Dot

as "unaroused" (meaning there was no minimal to no change in measured EDA) or "aroused" (meaning there were peaks) were quantified in relation to one another. For instance, if during all the moments when a youth was "soldering", 16 were labeled as "unaroused" and 42 were labeled as "aroused", the arousal ratio value would be 0.38 (derived from 16 divided by 42). If the numbers of unaroused and aroused moments were equal, the ratio would be 1.0. If there were more unaroused moments than aroused, then the ratio would be greater than 1.0. More details are in Cain and Lee (2020). The reason for computing unaroused as the numerator and aroused as the denominator was to reduce the number of undefined values because of division by 0. Since we were focused on what *was* engaging, we set aside activities that had no peaks (0 arousal moments).

As some illustrative excerpts of what this yielded, consider that when watching an adult mentor lecturing to the youth, this analysis showed that Dot had an arousal ratio of 0.26, whereas Jane had a ratio of 1.33. Dot had relatively more arousal moments in comparison to Jane, suggesting Dot had relatively more moments of situational engagement. On the other hand, when soldering, Jane had a ratio of 0.29 and Dot had 0.99. Thus, Jane appeared to have more instances of situational engagement than Dot. Both Dot and Jane had ratio values of 0 when they were presenting to or speaking in front of the larger group of maker camp attendees (with 8 and 13 aroused moments respectively).

Our core inference from this study of just the single pair is that even when doing the same activities together, there are some activities that seem, at least momentarily, more engaging for both youth and some that are more engaging for just one youth. In some respects, this is an obvious statement. In some ways, Dot and Jane were similar and in others, they were different. However, this study provided some empirical support and techniques with wearable EDA and camera instrumentation for inferring that Dot was more responsive when she was an observer of mentors. Jane was more responsive when directly engaging and manually involving herself in the activity. Both girls were responsive in situations that involved speaking to the larger group.

5 Study 2: EDA Referenced Engagement in Two Maker Camps

The second study that I completed with my team was with shorter multi-week activities (Lee et al., 2019) and involved a new analytical approach but was informed by findings from the first study. Several things differed across the studies that motivated exploration of new analytical approaches. In the first study, the makerspace program lasted for 12 weeks. In this second study, the programs were designated as "camps" (based on the registration system used by the makerspace) and lasted only 6 weeks. Each camp was advertised as focusing on the construction of a single artifact. In one camp, it was to make model rockets that would be launched in a community park. In the other camp, it was to make customized lanterns with media editing software and laser cutters. The lanterns would include carved imagery specific to each camper, based on their individual preferences.

Following from the first study where there did seem to be, at least for a pair of youth, some indications of common activities that we could interpret as triggering momentary engagement, we sought to determine what were activities that led to increases in EDA (peaks) for a large number of youth in the makerspace.

For this camp (rockets), we had 12 youth participating. In the second (lanterns), there were 13 youth (Fig. 13.3). We equipped each youth with E4 devices. Having recognized the importance of video rather than still image cameras to fully contextualize the activities taking place in the makerspace and from the perspective of focal youth, we opted to rely on wearable video cameras (GoPro Session cameras) that were worn with elastic chest mounts. Instead of reviewing what photographs were taken, we reviewed video clips. Additionally, we had a validated engagement survey instrument to use for potential concurrent validation (Bathgate & Schunn, 2017).

Given larger numbers of youth and differences in each individuals, we sought to find higher densities of EDA peaks for youth and where they aligned with one another in time. That is, we wanted to know what activity was taking place in the makerspace when at least a quarter of the enrolled youth had higher densities of

Fig. 13.3 Youth working with an adult in the Lantern maker camp

peaks relative to their session participation because we had reason from Study 1 to suspect that there would be few to no activities that were situationally engaging for all. This involved visual analysis of plots to identify timestamps and then subsequent review and coding of video footage for the primary activity. These were subject to reliability testing from multiple analysts and is described in detail in Lee et al. (2019).

Based on this second study, and looking across both the rocket and lantern camps, the makerspace activity that had the highest density of peaks across swaths of youth was peer socializing. The unstructured time when the youth could talk with one another about a multitude of topics that did not directly pertain to STEM or the materials at hand had more peaks. Following those, adult mentor-led instruction that was highly interactive – which had dialogic exchanges rather than just lecture – and physical making, when physical objects were being assembled, fabricated, or refined – were the next most peak-dense. The least frequent peak-dense activities included those that involved personal expression and freely seeking resources based on existing interests. By this, we mean that when youth were doing things like picking images from the internet to transfer to software for laser cutting or were painting rockets with the colors and decorations of their choice, there were some occasions where more periods of dense EDA peaks for multiple youth were present. However, the more social and active material activities were more frequently associated with these EDA responses.

Comparing with survey responses, we found moderate correlations with our peak detection approach and youth self-reported cognitive ($r = 0.645$, $p = 0.061$) and behavioral engagement ($r = 0.625$, $p = 0.072$) ratings. There was an insignificant relationship between the peak detection approach and self-reported affective ratings ($r = 0.330$, $p = 0.386$). Our interpretation of these findings is that EDA peaks seem to be related to cognitive effort and to active behavioral activity. The affective valence and the number of EDA peaks could not be determined. A period of time when a youth had more EDA peaks could be associated with positive feelings, negative feelings, or neutral feelings. Based on this analysis, it appears that some other approaches that go beyond EDA peak detection may be needed to gain information about affective valence in that dimension of engagement.

6 Study 3: EDA Referenced Engagement in an Extended Museum-Based Afterschool Maker Program

The third study of makerspace program EDA peak detection took place at a suburban makerspace at a large museum campus. This was a different physical site than the first two studies. The makerspace involved in this study hosted a weekly maker program each day of the week for different groups of local adolescent students who were registered. The youth were expected to attend regularly throughout the academic year for the same days (and thus, the same topics). The makerspace program that we followed was one that was focused on electronic textiles.

As discussed in Lee (2021), there were a number of relatable challenges that appeared with this program that are familiar to education researchers. These included inconsistent and varying youth registration and attendance at this discretionary program and staffing changes that led to changes in focus for the program over the course of the year.

However, despite those challenges, a group of four youth with high attendance who were consistently co-present were analyzed across 15 of the weekly sessions. These youth completed the same weekly engagement survey instrument, but of special interest in this analysis was their open-ended responses regarding what they would identify as the most "interesting" experience of the given day. "Interesting" was selected as the prompt as our early testing of prompts with youth suggested it was more comprehensible than "engaging".

Activities from study 2 and that were specific to this afterschool program (e.g., completing paper circuits, taking apart and rebuilding stuffed animals to practice sewing skills) were identified for analysis. In this analysis, we sought to determine who exhibited any EDA peak response during the activity. The associated open-response writings from the four focal youth for what they considered to be the most interesting were reviewed and compared against these EDA peak records.

First, we found that highly social conversational activities – specifically, planning and discussing future projects – had higher numbers of EDA peak responses across the four youth. On the other hand, the activity that had the lowest number of EDA peak response was when the youth were individually completing their sewing while watching videos on a shared TV in the makerspace. Based on live observations and video footage, there was little to no conversation or interaction across youth during this activity.

Second, the short answer written responses from youth were consistent with what the analysis of EDA said. One youth had a positive EDA response (peaks) during the paper circuits activity, and she commented about how learning how many different ways there were to complete a circuit was most interesting. Upon review of the video footage, the moment when this information was shared was while she was working on a paper circuit and this knowledge was being discussed and shared by the lead adult mentor near that youth. On a day that predominantly involved sewing while watching videos silently, one of the youth who gave a response to the question of what was interesting just responded that most of the day was "just" disassembling stuffed animals. That statement was broadly descriptive and did not make a statement about the activity being interesting or engaging. More details about quotes are available in Lee (2021). In short, while there were inherent challenges involved in completing empirical work with this setting and for this program, there were indications that EDA peaks, as we operationalized and restricted them, were suggestive of what youth found to be engaging. Moreover, the social interaction component again seems to be strongly associated with this measure of engagement.

7 Summary

Across multiple makerspace programs of different time durations, my team and I explored the use of relatively new unobtrusive wearable technologies to ascertain engagement. Specifically, these included EDA wearables and wearable cameras. Our focus was on engagement as a component and desired feature of learning experiences rather than on content knowledge, identity, or longer-term social and academic outcomes. This is an area of interest, particularly with respect to the maker movement and makerspaces in educational settings.

We advanced one approach for identifying a feature of EDA data – finding and counting peaks – and some criteria that we had established in ways that were intended to be conservative and specific to individuals. Our goal had been to go beyond the prior tradition of controlled laboratory settings with newly available measurement instruments that produce messy data that still are in search of broadly accepted standards for interpretation. I feel cautious about how firmly to draw definitive conclusions from this line of work, but I do feel comfortable lending qualified support for the following:

- EDA measurement in complex makerspaces does seem to produce some amount of information that is suggestive of increased youth engagement.
- Youth EDA responses are both similar and different. There are some activities that seem to elicit EDA responses from multiple youth. There are some activities that elicit EDA responses from just individual or small subsets of youth. This is sensible given that we should not expect engagement triggers to be universal. At

the same time, we should not expect triggers to be purely random or idiosyncratic to every single individual.

- Social activities, involving back-and-forth conversations and exchanges, that may not pertain to the maker topics at hand, seem to more elicit EDA response. In some ways this is not surprising – interacting with others should be engaging for many. Yet for makerspaces, where so much attention is directed to the fabrication equipment, the use of technology, and autonomous pursuit of individual interests, the social dynamics and interactions seem to be quite important.
- Maker activities that lack social interaction can be, for many youth, unengaging at least psychophysiologically. One can be doing maker work – sewing, assembling, customizing – and not exhibit a notable response, at least as far as EDA is concerned.

Having completed these multiple studies, and if pressed to offer recommendations for makerspaces and maker activities to educators and practitioners, then based on these years of person-in-context work I would offer the following: if you want to make this an engaging experience, make sure there are lots of times and spaces for the youth to talk to one another freely. While this may seem obvious, there are ways in which I could (and have) seen maker activities and spaces organized in ways that go against this recommendation. For example, some self-contained classrooms pursue maker activities but those are done in a very structured way where youth are to focus on the instructor and follow the model as given without interacting with one another. This would run counter to my recommendation and also the larger maker pedagogical philosophy. Another example is a makerspace that occupies a library, which may have norms of remaining quiet so as to not disturb other patrons. This would largely involve minimizing talk in order to maintain the quiet atmosphere. In other work with maker programs in libraries, I have anecdotally seen levels of enthusiasm and engagement from youth when they could freely talk, laugh, question, and opine to one another while completing a maker project. This appeared to be more important than what technology or project was involved.

Still, and as has been my tone throughout this chapter, I would not resolutely and definitively say this recommendation would apply unconditionally for maker learning experiences. My goal in this work has been to explore, experiment methodologically, and get some sense as to whether this approach to unobtrusively observing learning is worth continued inquiry in the future, whether by my own hand or by someone else's. On this point, I do feel more confident in asserting that it is worth continuing the exploration of wearable EDA as one source of data in spaces like these. They may not alone be sufficient to answer all of our questions about youth engagement in makerspaces, but they do seem to make some contributions that concur with what else we know from the extant literature and from what youth are willing to report back to us.

In considering the use of these wearables and the approaches described for unobtrusive observation of learning in digital environments, I would assert that this type of work is not without potential. It supports the generation of plausible, evidence-based inferences. However, it is new and not yet widespread, meaning more work and refinement on what kind of analysis to do with these instruments and approaches is needed. In disclosing what we have tried and why, the core hope is that those who are interested in these approaches take what we have attempted and do better still. It is unclear whether EDA can be informative alone, as so much of what triggered peaks are situational and some means of capturing and interpreting the complexity and breadth of the situation seems necessary. However, it does not seem to be a useless adjunct or data stream to include in observing learning. I do note that these claims should be understood as being most pertinent to studying learning in makerspaces. For work that involves unobtrusive observation of learning when learners are stationary, rather than moving freely and interacting in socially and physically complex spaces like a makerspace, other data streams and techniques, including those described in other chapters of this book should be given priority.

Acknowledgments Thanks to Ryan Cain, Liam Fischback, Diamond Deng, Aditya Chandel, and Kyle Lam for their assistance in the work reported here. This work was supported in part by funding from the National Science Foundation under Grant Nos. CNS-1623401 and CNS-1949740. The opinions expressed herein are those of the author and do not necessarily reflect those of the National Science Foundation.

References

Azevedo, R. (2015). Defining and measuring engagement and learning in science: Conceptual, theoretical, methodological, and analytical issues. *Educational Psychologist, 50*(1), 84–94. https://doi.org/10.1080/00461520.2015.1004069

Bathgate, M., & Schunn, C. (2017). Factors that deepen or attenuate decline of science utility value during the middle school years. *Contemporary Educational Psychology, 49*, 215–225. https://doi.org/10.1016/j.cedpsych.2017.02.005

Betancourt, M. A., Dethorne, L. S., Karahalios, K., & Kim, J. G. (2017). Skin conductance as an in situ marker for emotional arousal in children with neurodevelopmental communication impairments: Methodological considerations and clinical implications. *ACM Transactions on Accessible Computing, 9*(3), 8. https://doi.org/10.1145/3035536

Blikstein, P. (2013). Digital fabrication and 'making' in education: The democratization of invention. In J. Walter-Herrmann & C. Büching (Eds.), *FabLabs: Of machines, makers and inventors* (pp. 203–222). Transcript Publishers.

Cain, R., & Lee, V. R. (2020). Measuring electrodermal activity in an afterschool maker program to document engagement of a pair of students. In R. Zheng (Ed.), *Cognitive and affective perspectives on immersive technology in education* (pp. 128–150). IGI Global. https://doi.org/10.4018/978-1-6684-6295-9.ch026

Carrillo, E., Moya-Albiol, L., González-Bono, E., Salvador, A., Ricarte, J., & Gómez-Amor, J. (2001). Gender differences in cardiovascular and electrodermal responses to public speaking task: The role of anxiety and mood states. *International Journal of Psychophysiology, 42*(3), 253–264. https://doi.org/10.1016/S0167-8760(01)00147-7

Daily, S. B., James, M. T., Roy, T., & Darnell, S. S. (2015). EngageMe: Designing a visualization tool utilizing physiological feedback to support instruction. *Technology, Instruction, Cognition and Learning, 10*(2), 107–126.

Dawson, M. E., Schell, A. M., & Filion, D. L. (2007). The electrodermal system. In J. T. Cacioppo, L. G. Tassinary, & G. G. Berntson (Eds.), *Handbook of psychophysiology* (pp. 159–181). Cambridge University Press.

Di Lascio, E., Gashi, S., & Santini, S. (2018). Unobtrusive assessment of students' emotional engagement during lectures using electrodermal activity sensors. *Proceedings of the ACM on Interactive, Mobile, Wearable and Ubiquitous Technologies, 2*(3), 1–21. https://doi.org/10.1145/3264913

Dougherty, D. (2013). The maker mindset. In M. Honey & D. Kanter (Eds.), *Design, make, play: Growing the next generation of STEM innovators* (pp. 7–11). Taylor & Francis.

Dweck, C. S. (2006). *Mindset: The new psychology of success*. Random House.

Eisenhauer, S. (2019). Youths' individual pathways towards contextual well-being: Utilizing electrodermal activity as an ethnographic tool at a theater after-school program. *Ethos, 47*(2), 168–189. https://doi.org/10.1111/etho.12235

Fields, D. A., & Lee, V. R. (2016). Craft Technologies 101: Bringing making to higher education. In K. Peppler, E. Halverson, & Y. Kafai (Eds.), *Makeology* (Vol. 1, pp. 121–137). Routledge.

Fredricks, J. A., Blumenfeld, P. C., & Paris, A. H. (2004). School engagement: Potential of the concept, state of the evidence. *Review of Educational Research, 74*(1), 59–109. https://doi.org/10.3102/00346543074001059

Gashi, S., Lascio, E. D., & Santini, S. (2019). Using unobtrusive wearable sensors to measure the physiological synchrony between presenters and audience members. *Proceedings of the ACM on Interactive, Mobile, Wearable and Ubiquitous Technologies, 3*(1), Article 13. https://doi.org/10.1145/3314400

Gruzelier, J. H., & Venables, P. H. (1972). Skin conductance orienting activity in a heterogeneous sample of schizophrenics: Possible evidence of limbic dysfunction. *The Journal of Nervous and Mental Disease, 155*(4), 277–287. https://doi.org/10.1097/00005053-197210000-00007

Hernandez, J., Riobo, I., Rozga, A., Abowd, G. D., & Picard, R. W. (2014). Using electrodermal activity to recognize ease of engagement in children during social interactions. In *Proceedings of the 2014 ACM international joint conference on pervasive and ubiquitous computing, Seattle, Washington, DC*. https://doi-org.stanford.idm.oclc.org/10.1145/2632048.2636065

Hidi, S., & Renninger, K. A. (2006). The four-phase model of interest development. *Educational Psychologist, 41*(2), 111–127. https://doi.org/10.1207/s15326985ep4102_4

Kuhn, T. S. (1962). *The structure of scientific revolutions*. University of Chicago Press.

Lal, S., Eysink, T. H., Gijlers, H. A., Verwey, W. B., & Veldkamp, B. P. (2021). Detecting emotions in a learning environment: A multimodal exploration. In *Proceedings of EC-TEL (Doctoral consortium)*.

Lee, V. R. (2019). On researching activity tracking to support learning: A retrospective. *Information and Learning Sciences, 120*(1/2), 133–154. https://doi.org/10.1108/ILS-06-2018-0048

Lee, V. R. (2021). Youth engagement during making: using electrodermal activity data and first-person video to generate evidence-based conjectures. *Information and Learning Sciences, 122*(3/4), 270–291. https://doi.org/10.1108/ILS-08-2020-0178

Lee, V. R., & Shapiro, R. B. (2019). A broad view of wearables as learning technologies: Current and emerging applications. In P. Diaz, A. Ioannou, K. K. Bhagat, & J. M. Spector (Eds.), *Learning in a digital world – perspectives on interactive technologies for formal and informal education* (pp. 113–133). Springer. https://doi.org/10.1007/978-981-13-8265-9_6

Lee, V. R., Fischback, L., & Cain, R. (2019). A wearables-based approach to detect and identify momentary engagement in afterschool Makerspace programs. *Contemporary Educational Psychology, 59*. https://doi.org/10.1016/j.cedpsych.2019.101789

Martin, L. (2015). The promise of the Maker Movement for education. *Journal of Pre-College Engineering Education Research (J-PEER), 5*(1), 4. https://doi.org/10.7771/2157-9288.1099

Melo, M., & Nichols, J. (Eds.). (2020). *Re-making the library makerspace: Critical theories, reflections, and practices*. Library Juice Press.

Menghini, L., Gianfranchi, E., Cellini, N., Patron, E., Tagliabue, M., & Sarlo, M. (2019). Stressing the accuracy: Wrist-worn wearable sensor validation over different conditions. *Psychophysiology, 56*(11), e13441. https://doi.org/10.1111/psyp.13441

Miller, B. W. (2015). Using reading times and eye-movements to measure cognitive engagement. *Educational Psychologist, 50*(1), 31–42. https://doi.org/10.1080/00461520.2015.1004068

Milstein, N., & Gordon, I. (2020). Validating measures of electrodermal activity and heart rate variability derived from the Empatica E4 utilized in research settings that involve interactive dyadic states. *Frontiers in Behavioral Neuroscience, 14*, Article 148. https://doi.org/10.3389/fnbeh.2020.00148

Naveteur, J., Freixa, I., & Baque, E. (1987). Individual differences in electrodermal activity as a function of subjects' anxiety. *Personality and Individual Differences, 8*(5), 615–626. https://doi.org/10.1016/0191-8869(87)90059-6

Papert, S. (1980). *Mindstorms: Children, computers, and powerful ideas*. Basic Books.

Peppler, K., & Glosson, D. (2013). Stitching circuits: Learning about circuitry through e-textile materials. *Journal of Science Education and Technology, 22*(5), 751–763. https://doi.org/10.1007/s10956-012-9428-2

Peppler, K., Halverson, E. R., & Kafai, Y. B. (2016a). *Makeology: Makers as learners* (Vol. 2). Routledge.

Peppler, K., Halverson, E., & Kafai, Y. B. (2016b). *Makeology: Makerspaces as learning environments* (Vol. 1). Routledge.

Poh, M.-Z., Swenson, N. C., & Picard, R. W. (2010). A wearable sensor for unobtrusive, long-term assessment of electrodermal activity. *IEEE Transactions on Biomedical Engineering, 57*(5), 1243–1252. https://doi.org/10.1109/tbme.2009.2038487

Renninger, K. A., & Bachrach, J. E. (2015). Studying triggers for interest and engagement using observational methods. *Educational Psychologist, 50*(1), 58–69. https://doi.org/10.1080/00461520.2014.999920

Setz, C., Arnrich, B., Schumm, J., Marca, R. L., Trster, G., & Ehlert, U. (2010). Discriminating stress from cognitive load using a wearable EDA device. *IEEE Transactions on Information Technology in Biomedicine: A Publication of the IEEE Engineering in Medicine and Biology Society, 14*(2), 410–417. https://doi.org/10.1109/titb.2009.2036164

Sheridan, K., Halverson, E. R., Litts, B., Brahms, L., Jacobs-Priebe, L., & Owens, T. (2014). Learning in the making: A comparative case study of three makerspaces. *Harvard Educational Review, 84*(4), 505–531. http://www.metapress.com/content/BRR34733723J648U

Sinatra, G. M., Heddy, B. C., & Lombardi, D. (2015). The challenges of defining and measuring student engagement in science. *Educational Psychologist, 50*(1), 1–13. https://doi.org/10.1080/00461520.2014.1002924

Soltis, N. A., McNeal, K. S., Atkins, R. M., & Maudlin, L. C. (2020). A novel approach to measuring student engagement while using an augmented reality sandbox. *Journal of Geography in Higher Education, 44*(4), 512–531. https://doi.org/10.1080/03098265.2020.1771547

Taylor, S., Jaques, N., Chen, W., Fedor, S., Sano, A., & Picard, R. (2015). Automatic identification of artifacts in electrodermal activity data. In *International conference of the IEEE Engineering in Medicine and Biology Society (EMBC)*. IEEE. https://doi.org/10.1109/EMBC.2015.7318762

Vossoughi, S., Hooper, P. K., & Escudé, M. (2016). Making through the lens of culture and power: Toward transformative visions for educational equity. *Harvard Educational Review, 86*(2), 206–232. https://doi.org/10.17763/0017-8055.86.2.206

Worsley, M., & Blikstein, P. (2018). A multimodal analysis of making. *International Journal of Artificial Intelligence in Education, 28*(3), 385–419. https://doi.org/10.1007/s40593-017-0160-1

Xie, K., Heddy, B. C., & Vongkulluksn, V. W. (2019). Examining engagement in context using experience-sampling method with mobile technology. *Contemporary Educational Psychology, 59*, 101788. https://doi.org/10.1016/j.cedpsych.2019.101788

Zhang, J., Wang, K., & Zhang, Y. (2021). Physiological characterization of student engagement in the naturalistic classroom: A mixed-methods approach. *Mind, Brain, and Education, 15*(4), 322–343. https://doi.org/10.1111/mbe.12300

Chapter 14
Collecting Unobtrusive Data: What Are the Current Challenges?

Roberto Martinez-Maldonado

Abstract Sensing technologies are rapidly dropping in price and improving the quality of data acquisition. It is therefore expected that sensing technologies, paired with artificial intelligence algorithms, will become a common part of the educational researcher's toolkit to unobtrusively measure learning phenomena in years to come. In this section, we learned about the potential of using multimodal and multichannel data to create rich models of higher-order constructs, namely engagement, self-regulated learning (SRL) and collaboration. The five chapters showcased various applications of sensing technologies and logging mechanisms to generate indicators that can be critical for studying and supporting learning.

Keywords Unobtrusive observation · Data collection · Self-regulated learning · Collaboration · Multimodal data

1 A Critical Overview of the Chapters

In Chap. 8, Prahraj et al. described some critical steps towards automating the generation of collaboration indicators based on audio data, making inroads into designing end-user interfaces that teachers and students could use. The provision of end-user multimodal analytics interfaces is rare, partly because of the complexity of transforming low-level data into meaningful information. Prahraj et al.

R. Martinez-Maldonado (✉)
Monash University, Clayton, Australia
e-mail: Roberto.MartinezMaldonado@monash.edu

V. Kovanovic et al. (eds.), *Unobtrusive Observations of Learning in Digital Environments*, Advances in Analytics for Learning and Teaching,
https://doi.org/10.1007/978-3-031-30992-2_14

demonstrated how this could be done for the case of microphone data by cleaning less important words (e.g., stop words) and distilling keywords that can indicate topics that may mean something to educational stakeholders.

As sensor data is increasingly being used in educational research, a much-needed discussion about the challenges of extracting meaningful information from sensor data, such as electrodermal activity (EDA) data, has been provided by Lee in Chap. 9. Most of the EDA sensors used in educational research have not (so far) been designed for highly dynamic educational activities such as those that commonly occur in maker spaces. Therefore, it is crucial to recognise that sensor data is commonly noisy and incomplete. Deriving indicators from such fuzzy data sources requires lots of exploration, like the one described in Lee's chapter. To provide context, Lee explored the use of wearable physiological wristbands and wearable still image cameras to enrich the engagement analysis in a physical learning space and complement the sensor data.

As shown by Salehian Kia et al. in Chap. 10, multichannel data itself could serve researchers to establish the validity of the mapping from low-level trace data to higher-order constructs, such as the phases of the SRL process. If the same learning event can be observed on multiple data channels, the data streams can validate one another or add meaning to low-level data. This approach can help researchers perform a deeper analysis of logged activity happening in various digital spaces (e.g., at the learning management system, an assessment platform and a digital textbook) and provide effective interventions in the future.

With the growing amount and diversity of educational data, it is critical to ensure that the use of such data is grounded in sound educational theories. In Chap. 11, Wiedbusch et al. provided a theoretically grounded approach for measuring engagement with multimodal data originating from self-reports, log streams, oculometrics, physiological sensors, facial expressions, body gestures, think-aloud and emote-alouds. Their goal is to measure SRL engagement by considering all the aspects of students, including what happens inside their heads as well as emotional and social aspects that are often hard to inspect with the 'naked eye'. Authors envision a set-up where multiple sensors, computer vision algorithms and educational tools are part of a synchronised ecosystem capable of recognising the behavioural, cognitive, emotional and agentic engagement states of learners to provide some specific support while students work in front of the computer.

In this regard, in Chap. 12, Malmberg et al. also discussed their stance on using multimodal data to study engagement from a collaborative learning perspective as students experience socio-emotional interactions. Like Lee, the authors also use EDA sensors, but this time, to analyse how physiological synchrony can aid in understanding the relationship between cognitive, socio-emotional and interaction episodes in group-level regulation. The authors illustrate how multimodal data can augment conversation analysis which has been a staple technique used to study collaborative learning.

2 Using Multimodal Data for Unobtrusive Measurement of Learning: Where Are We Now?

Overall, using multimodal data to unobtrusively measure learning can be seen as an emerging and exciting area still in its infancy. One of the key challenges that researchers in this area often face concerns the ecological *validity of indicators* and the process of imbuing the data captured from sensors with meaningful constructs relevant to a specific learning context for educational stakeholders to make sense of resulting algorithmic outputs (Cukurova et al., 2020; Yan et al., 2022; Di Mitri et al., 2018). We can minimise noisy data if we conduct controlled experiments and take the time to carefully analyse and manually fill the gaps in multimodal data. However, eventually, we want to close the analytics loop by supporting teachers and students where learning happens. Lab studies will continue to be critical for investigating the validity of specific indicators and advancing educational research (Sharma & Giannakos, 2020). However, to make a practical impact, multimodal studies need to make it into the actual classroom, where multiple confounding variables can be introduced as they reflect the authentic (often 'messy') conditions in which learning ultimately occurs (Worsley et al., 2021). Chejara et al. (2023) and Martinez-Maldonado et al. (2023) have recently discussed the several practical, logistic and ethical challenges that can be identified only when multimodal technologies are deployed *in-the-wild,* which can strongly shape the ultimate effectiveness of multimodal data-enhanced educational interventions. Nonetheless, enriching authentic learning spaces with multimodal analytics and data sensing capabilities can hold the potential to help researchers study the umbrella of expected and unexpected events that can shape learning.

As multimodal data can help us create a richer picture of the learning activity, it can also increase the *complexity of the potential pedagogical intervention.* This points to a second key challenge: how can we create fully automated multimodal tools that provide some direct benefit to teachers and students? To address this challenge, there is a need to design end-user interfaces more carefully to help audiences who usually are not formally trained in data analysis gain insight into multimodal educational data. In several studies where multimodal data is used, humans still need to be part of the pre-processing data process (i.e., see review by Praharaj et al., 2021). While this adds validity and rigour to a study from an educational research standpoint, it also makes innovation *in the wild* more challenging. Fully automating the whole analytics process, from multimodal or multichannel data acquisition to creating interfaces that are 'easy to use' and meaningful to end-users in the educational sector, requires a multidisciplinary team of experts in data science, human-computer interaction and education (Yan et al., 2022). Unfortunately, not all research centres have the resources to form such multidisciplinary teams. As a result, the collaboration between researchers and practitioners in multimodal learning analytics is crucial for advancing the field and keeping it thriving.

A third challenge illustrated across these previous chapters is related to the *critical role of the educational context* in giving meaning to multimodal data. While one data channel can help provide context to another channel, the ultimate meaning of any indicator extracted from data depends on the learning task and, hence, on the pedagogical intentions of the learning design (Ochoa, 2022). For example, the detection of the quality of collaboration is highly contextual. Thus, collaboration indicators can be identified based on the learning goals (the learning design) and using educational and teamwork theories. Theoretical constructs, such as those found in SRL and collaborative learning theory, can give meaning to the indicators obtained from fuzzy physiological data (e.g., Azevedo et al., 2022). Ultimately, education is highly contextual. Therefore, it is not expected to treat multimodal learning analytics innovations as one type of solution that can be applied to multiple contexts but as an approach for embracing the complexity and particularities of each educational context. These multimodal innovations also highlight the limitations of just analysing the clickstreams and keystrokes that students perform in the learning management system by considering the broader context of using the socio-technical context where learning happens (Echeverria et al., 2019).

These and other challenges in multimodal learning analytics research can be seen as opportunities yet to be explored. In any case, the current technical limitations in sensing technologies and analysis approaches are being addressed by the rapid progress of artificial intelligence (AI). For example, the automated transcription problem that previously hindered educational researchers from developing fully automated tools to aid face-to-face collaboration is now close to being entirely resolved (Southwell et al., 2022). Improvements in human voice detection, noise filtering, speaker diarisation and automated transcription algorithms generate conversation logs similar to those of professional human transcription services. The physiological wristbands currently used in educational research, not specifically created for educational purposes, will soon be replaced by better sensors less susceptible to the physical activity and ambient conditions found in most classrooms. Advances in ubiquitous and pervasive computing will only further augment our capacity to gather more and more data. Hence, it will be critical to further advance approaches for extracting meaning from multimodal data while also considering the ethical implications of using such data.

Indeed, a topic that has not been deeply covered in the chapters presented in this section involves the ethical and privacy implications of unobtrusively gathering multimodal sensor and log data from students. Just because we can capture more data does not mean we should do it. If we do it, much more discussion about who owns these data is required. The danger of excessive surveillance is genuine, and it is still difficult to predict all the possible scenarios concerning what education providers can do with such detailed and frequently sensitive data. What limitations will be placed on the utilisation of these data? Most importantly, if the aim is to create more accurate learning models, what will happen for students that can be considered 'outliers' from a data science perspective but may be demonstrating unique learning pathways from a social science perspective? Finally, there is a pressing need for more dialogue on how to involve teachers, students and other educational stakeholders in the design process of multimodal learning analytics (Echeverria et al., 2022). Several important decisions are taken during the design process of any programmable tool. How can the values of educational stakeholders be considered in the design process? This calls for a human-centred design perspective that provides a channel for relevant stakeholders easily affected by data-intensive initiatives to voice their concerns and remain active agents in the learning process rather than passive receivers. Since AI and sensing technologies are already impacting educational practices, we must create systems that exploit these technologies with integrity.

References

Azevedo, R., Bouchet, F., Duffy, M., Harley, J., Taub, M., Trevors, G., Cloude, E., et al. (2022). Lessons learned and future directions of metatutor: Leveraging multichannel data to scaffold self-regulated learning with an intelligent tutoring system. *Frontiers in Psychology, 13.*

Chejara, P, Prieto, L. P., Rodríguez-Triana, M. J., Ruiz-Calleja, A., Kasepalu, R. & Shankar, S. K. (2023). Multimodal Learning Analytics research in the wild: challenges and their potential solutions. In *LAK 2023 workshop: Leveraging multimodal data for generating meaningful feedback* (pp. 1–5).

Cukurova, M., Giannakos, M., & Martinez-Maldonado, R. (2020). The promise and challenges of multimodal learning analytics. *British Journal of Educational Technology, 51*(5), 1441–1449.

Di Mitri, D., Schneider, J., Specht, M., & Drachsler, H. (2018). From signals to knowledge: A conceptual model for multimodal learning analytics. *Journal of Computer Assisted Learning, 34*(4), 338–349.

Echeverria, V., Martinez-Maldonado, R., & Buckingham Shum, S. (2019). Towards collaboration translucence: Giving meaning to multimodal group data. In *Proceedings of the 2019 SIGCHI conference on human factors in computing systems* (pp. 1–16).

Echeverria, V., Martinez-Maldonado, R., Yan, L., Zhao, L., Fernandez-Nieto, G., Gašević, D., & Shum, S. B. (2022). HuCETA: A framework for human-centered embodied teamwork analytics. *IEEE Pervasive Computing*, 1–11.

Martinez-Maldonado, R., Echeverria, V., Fernandez-Nieto, G., Yan, L., Zhao, L., Alfredo, R., Li, X., Dix, S., Jaggard, H., Wotherspoon, R., Osborne, A., Gasevic, D., & Buckingham Shum, S. (2023). Lessons learnt from a multimodal learning analytics deployment in-the-wild. *ACM Transactions on Computer-Human Interaction*. In press.

Ochoa, X. (2022). Multimodal learning analytics – Rationale, process, examples, and direction. In C. Lang, G. Siemens, A. F. Wise, D. Gašević, & A. Merceron (Eds.), *The handbook of learning analytics* (pp. 54–65). SOLAR.

Praharaj, S., Scheffel, M., Drachsler, H., & Specht, M. (2021). Literature review on co-located collaboration modeling using multimodal learning analytics—Can we go the whole nine yards? *IEEE Transactions on Learning Technologies, 14*(3), 367–385.

Sharma, K., & Giannakos, M. (2020). Multimodal data capabilities for learning: What can multimodal data tell us about learning? *British Journal of Educational Technology, 51*(5), 1450–1484.

Southwell, R., Pugh, S., Perkoff, E. M., Clevenger, C., Bush, J. B., Lieber, R., Ward, W., Foltz, P. & D'Mello, S. (2022). Challenges and feasibility of automatic speech recognition for modeling student collaborative discourse in classrooms. In *Proceedings of the 15th international conference on educational data mining* (pp. 302–315).

Worsley, M., Martinez-Maldonado, R., & D'Angelo, C. (2021). A new era in multimodal learning analytics: Twelve core commitments to ground and grow MMLA. *Journal of Learning Analytics, 8*(3), 10–27.

Yan, L., Zhao, L., Gasevic, D., & Martinez-Maldonado, R. (2022). Scalability, sustainability, and ethicality of multimodal learning analytics. In *LAK22: 12th international learning analytics and knowledge conference* (pp. 13–23).

Correction to: Measuring and Validating Assumptions About Self-Regulated Learning with Multimodal Data

Fatemeh Salehian Kia, Mathew L. Bernacki, and Jeffrey A. Greene

Correction to:
Chapter 9 in: V. Kovanovic et al. (eds.), *Unobtrusive Observations of Learning in Digital Environments*, Advances in Analytics for Learning and Teaching, https://doi.org/10.1007/978-3-031-30992-2_9

This book was inadvertently published with Dr. Greene's first name misspelt as Jeffery and is now corrected to reflect his name as Jeffrey.

The updated original version of this chapter can be found at https://doi.org/10.1007/978-3-031-30992-2_9

V. Kovanovic et al. (eds.), *Unobtrusive Observations of Learning in Digital Environments*, Advances in Analytics for Learning and Teaching, https://doi.org/10.1007/978-3-031-30992-2_15

Index

V. Kovanovic et al. (eds.), *Unobtrusive Observations of Learning in Digital Environments*, Advances in Analytics for Learning and Teaching,
https://doi.org/10.1007/978-3-031-30992-2

MIX
Papier aus verantwortungsvollen Quellen
Paper from responsible sources
FSC® C105338

If you have any concerns about our products,
you can contact us on
ProductSafety@springernature.com

In case Publisher is established outside the EU,
the EU authorized representative is:
Springer Nature Customer Service Center GmbH
Europaplatz 3, 69115 Heidelberg, Germany

Printed by Libri Plureos GmbH
in Hamburg, Germany